CW00342137
218
54
204
164
134
120
74
24
154
154

THE ATLAS OF UNUSUAL BORDERS

ZORAN NIKOLIĆ

CONTENTS

INTRODUCTION
(UN)USUAL GEOGRAPHY

Even as a young child, one glance at a map was enough to grab my attention and excite my curiosity. I wondered what those colourful areas and lines were, what it meant that we were living 'here', and what boundaries were at all. As I grew older and learnt some basic geography, the map became, for me, an even more wonderful invention, like a travelling machine that allowed me to just pop over from India to Argentina, then from there to Australia and immediately after that to Greenland. Gradually I discovered more and more interesting details on my maps and atlases, including some 'new' countries and unusual borders. Although my educational path led me away to economics and computer science, it was clear that geography would forever remain my favourite science.

But what actually is geography? It is a complex subject, which studies the natural and social phenomena on Earth. Its name derives from the Greek words γεω (geo, meaning 'Earth') and γραφία (graphia, meaning 'description'), so this science could be translated as 'description of the Earth'.

The complexity of geography is reflected in the fact that it is a bridge between the natural and social sciences. Roughly speaking, physical geography is concerned with the overall nature of the Earth and its spheres: atmosphere, lithosphere, biosphere and hydrosphere; while social geography studies population, economy and settlements.

While studying the natural or social aspects of our Blue Planet, the geographer will encounter some curious phenomena. These may either be really big or really small natural features; or phenomena that occur at only a few locations; or they may be unusual or illogical borders. It would be impossible to cover all of the Earth's oddities and peculiarities in just one book, but here are presented some of what are, I hope, the most interesting and unusual ones.

By a combination of circumstances, I was born in 1975 in Gornji Milanovac, central Serbia. I only spent the first two years of my life there, since my family relocated a lot, within the vicinity of Belgrade. These relocations probably influenced me as a child, because at some point in my childhood I took a map of what was then Yugoslavia and tried to find every place where we had lived. And so began my fascination with geography, maps and atlases, a fascination that has lasted until now, when paper maps are almost a thing of the past. I have nothing against digital atlases – on the contrary, I consider them an excellent way of modernizing the representation of geographical objects. However, in my house at all times there must be at least one solid, printed geographic atlas, even if I only glance at it once a year. Although my diploma states that I am an IT engineer, geography and maps have never stopped occupying my attention, and online atlases have become some of my favourite websites. So much time spent browsing and analyzing different maps has led me to notice a lot of unusual things, which has provided the inspiration to write this book and to try to present, in one place, as many of these strange phenomena as possible.

I hope this book will be an interesting addition to all those – amateurs, professionals or students – who study geography, cartography, politics, and society. The book is written in such a way that it may be read in parts, so that everyone can find something of interest, while other chapters could just be skimmed through.

Zoran Nikolić

UNUSUAL BORDERS:
ENCLAVES, EXCLAVES, AND OTHER PHENOMENA

Borders are scars on the face of the planet...
Las fronteras dividen, solo crean cicatrices

These two lines are taken from the song 'We Rise Again' by the American band Gogol Bordello. The second line translates as: Borders divide, and only create scars.

Since the beginning of modern civilization, man has sought to claim his territory and define it with **borders**. These borders divide the land into what is 'mine' and what is 'yours'. They say that you need permission to see his mountain; that they must ask before watching the sunset on your beach; that this side of the lake is mine, and the opposite side is yours.

Borders do not always 'behave' in a predictable way. They do not always follow a simple, straight course, with as few twists and turns as possible. Where borders are not clearly defined, this creates potential for conflict. But perhaps the main cause of conflict is man's instinct to strive for more. Over many years, conflicts – both large and small – have led to the movement of borders, as people have sought to make the dream come true that 'ours' is greater than 'theirs'.

The political map of the world reveals the results of these border movements, the lines giving the impression of scars left by man on the face of our planet, and the course that they follow seeming, in some cases, to be illogical.

Enclaves (territories completely surrounded by the territory of another state) and **exclaves** (part of a territory or state that can be reached from its home territory only through another territory or state) are examples of such oddities. Schematically, we can put it this way:

In this case, C is an exclave of territory B, but at the same time it is an enclave within territory A.

Apart from enclaves, often known as true enclaves, there are also so-called **pene-enclaves** – territories that are physically separated from the home country but which can still be reached without passing through another country. An example of this is Alaska: although it can only be reached over land from the rest of the USA by travelling through Canada, it can be reached by sea without passing through foreign territorial waters. Perhaps surprisingly, there are many such localities around the world, and some are even more complicated.

MEĐURJEČJE

BOSNIA AND HERZEGOVINA | SERBIA

The wedding gift that became an enclave

43º 33' 38"N | 19º 25' 10"E

NIKOLIĆI

RUDO

REPUBLIKA SRPSKA
(BOSNIA AND HERZEGOVINA)

MIOČE

USTIBAR

PRIBOJ

MEĐURJEČJE
(BOSNIA AND
HERZEGOVINA)

SASTAVCI

SERBIA

0 2 km

Only fifteen kilometres from the tri-border area between Serbia, Bosnia and Herzegovina and Montenegro, lies the small village of **Međurječje**. Though similar to all the other villages in this region, Međurječje has one important characteristic that distinguishes it from the rest: it belongs to Bosnia and Herzegovina, though it is surrounded entirely by the territory of Serbia.

The reason for this – according to a popular local story – is that during the time of the Ottoman Empire, a Bosnian *Bey* gifted one of his wives some 400 hectares of land and woods in the vicinity of Priboj in Serbia. When demarcation was established between Austria-Hungary and Turkey, this land was annexed to Bosnia, which was occupied at that time by the Habsburg Empire. Ever since then, this land has been classed as part of the municipality of Rudo, in Bosnia and Herzegovina. In the meantime, the village of Međurječje was built there.

While the Yugoslav states were united as one country, it was of little importance which piece of land belonged to which republic or province, so it was the local authority of Priboj, Serbia, that provided the infrastructure – roads, power supply, school, police station, etc. – for the village of Međurječje. This has created the situation whereby the inhabitants pay their taxes to the authorities of Rudo (Bosnia and Herzegovina, or more specifically the sub-division of Republika Srpska), while all the other utility bills are paid to Priboj (Serbia).

The exclave, covering an area of a little under 400 hectares (equivalent to about 550 football fields), lies just over a kilometre from Republika Srpska. Most of the 270 inhabitants have either Serbian or dual citizenship.

Children from Međurječje, along with those from nearby villages in Priboj, attend classes in the primary school which, although located within this Bosnian exclave, follows the curriculum of Serbia.

An additional peculiarity is the fact that the local council offices of the Serbian village of **Sastavci** are located within the village of Međurječje, while this whole enclave is administered by the Rudo local council offices in Mioče. The village of Sastavci is located in Serbia, along the very border of this enclave. Due to their proximity and unclear borders, the towns of Međurječje and Sastavci in many ways function as one settlement, even though a state border officially divides them.

It is interesting to note that Međurječje is also surrounded by a religious border: the enclave is part of the Metropolitanate of Dabar-Bosnia of the Serbian Orthodox Church, while the surrounding villages are part of the Eparchy of Mileševa.

An important regional road passes through the enclave, creating problems for the inhabitants of some of the villages within the Priboj municipality when travelling to their administrative centre at Priboj town. The solution to this problem has yet to be agreed on: Serbia proposes an exchange of territories with Bosnia and Herzegovina, in order to 'correct' the border, while Bosnia believes that establishing a corridor between Rudo and Međurječje would be a better solution.

LLÍVIA
FRANCE | SPAIN

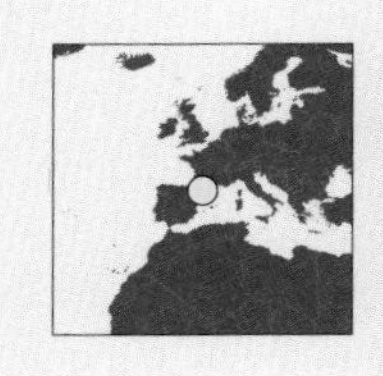

The place that became an enclave because it had been given town status

42º 28' 08"N | 1º 58' 48"E

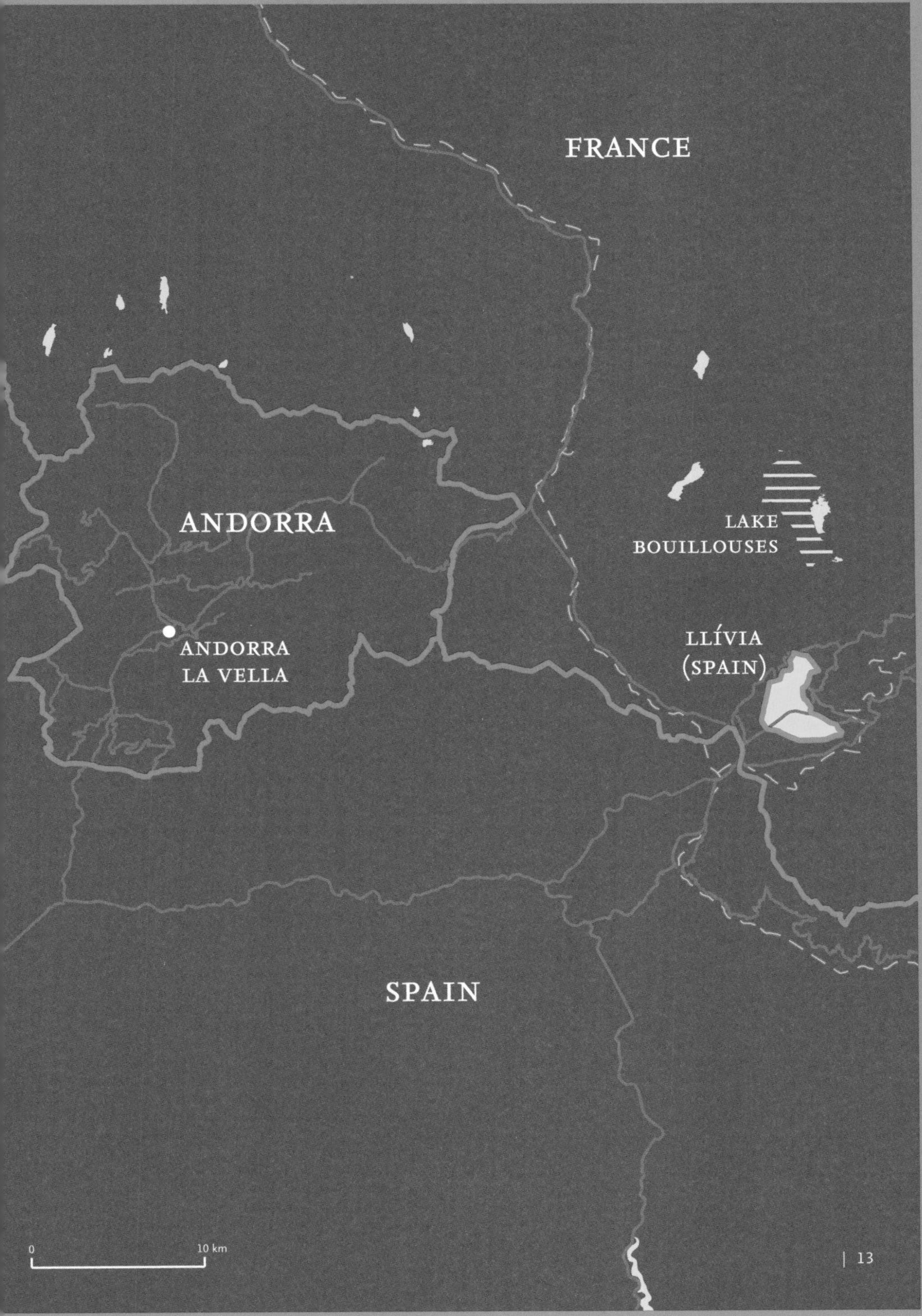

FRANCE

ANDORRA

LAKE
BOUILLOUSES

LLÍVIA
(SPAIN)

ANDORRA
LA VELLA

SPAIN

0 10 km

Llívia is a Spanish (or, more precisely, Catalan) town, within the territory of France. It is located about twenty kilometres east of Andorra and one kilometre from the Spanish–French border. Approximately 1,500 inhabitants live within the twelve square kilometres of this Spanish enclave in France.

Llívia became an enclave in an interesting way. A significant factor was the granting of town status upon Llívia as far back as the early Middle Ages. It had been the ancient capital of one of the Catalan countries, Cerdanya, and played a key role in the second half of the seventeenth century when Spain and France established their border through the Pyrenees. According to the border agreement, Spain was obliged to hand over all of the villages of Northern Cerdanya to France. And so it did. However, as Llívia had the status of a town, Spain retained this detached piece of land.

The inhabitants of Llívia often regard their town as the 'birthplace' of Catalonia, because their medieval ruler, Count Sunifred, ruling from Llívia itself, laid the foundations of the present-day Catalan identity.

In order to make up for the fact that Llívia had become detached from its homeland, it was allocated a relatively large area of land (hatched on the map) along the western side of Lake Bouillouses, north of the town. Although this land belongs to Llívia, it remains under French state sovereignty. Today, the crystal-clear lake, lying at about 2000 metres above sea level, is a popular holiday destination. Apart from tourism, it is important for dairy farming and the production of high-quality milk.

Europe's oldest pharmacy, founded in the early fifteenth century, is located in Llívia. It was donated to the town and made into a museum of pharmacy. The permanent display contains a large collection of original blue earthenware jars (albarello), which were used to store ointments and dried remedies in medieval pharmacies. Among the exhibits there is a large quantity of old medicines and cosmetic products, as well as one of the largest collections of prescriptions in Europe.

Despite the rural character of Llívia, this place represents a significant tourist attraction, not only because of its status as an enclave. An important music festival is held there every year, which gathers together the most well-known musicians from across the Cerdanya region. The Pharmacy Museum also attracts a large number of visitors, as well as the ruins of a fortress on the nearby hill.

BREZOVICA

CROATIA | SLOVENIA

It seems that one of the houses belongs to neither country

45° 41' 27"N | 15° 18' 10"E

CROATIA

BREZOVICA
PRI METLIKI

BREZOVICA
ŽUMBERAČKA
(CROATIA)

SLOVENIA

0 250 m

When a country is divided up, it often follows that what was a complicated internal border becomes an equally complicated international border. Such was the case along one section of the Slovenian–Croatian border, not far from the Croatian town of Karlovac. During the time of Yugoslavia, the federal border between Slovenia and Croatia – constituent elements of the former country – passed through the village of **Brezovica.** The largest part of this village, **Brezovica pri Metliki**, was on the Slovenian side of the border and the smaller part, **Brezovica Žumberačka**, belonged to the Croatian municipality of Ozalj, now within Karlovac County.

Brezovica Žumberačka has only a few houses, with thirty or so inhabitants, occupying an area of less than two hectares, but with the 'neighbouring' Slovenian village, Brezovica pri Metliki, it forms one place. Interestingly, it seems that the Croatian and Slovenian authorities are not entirely sure exactly where the border line is. It is even possible that there are currently a few more miniature enclaves and exclaves. Although this situation is less of an issue now that Slovenia and Croatia have entered into the European Union, a bizarre possibility is the fact that one house, together with the land around it, does not belong to either country. This would make it the so-called terra nullius, namely, no man's land. This has created an opportunity for the proclamation of an independent country, which was exploited in a virtual way. A website for the newly formed Kingdom of Enclava emerged on the Internet, though it had nothing to do with the inhabitants of the house itself. After the Slovenian government officially declared it to be their territory, the Enclava moved to one of the disputed islands in the River Danube, on the border between Serbia and Croatia.

Also in this part of the world, a hundred kilometres or so northeast of Brezovica, there is a possible triple tri-border area of Slovenia, Croatia and Hungary. According to some maps, a small part of Slovenia on the Mura River appears to be separate from the rest of the country and is located between Hungary and Croatia. Because it lies between two countries, this unnamed territory is not a true enclave, but it is certainly a Slovenian exclave, with two of its own tri-borders. Since this part of the border has not yet been fully defined, there may yet be changes to it in the future.

CAMPIONE D'ITALIA

ITALY | SWITZERLAND

Lying less than a kilometre from its home country, yet nearly fifteen kilometres by road

45° 58' 34"N | 08° 58' 28"E

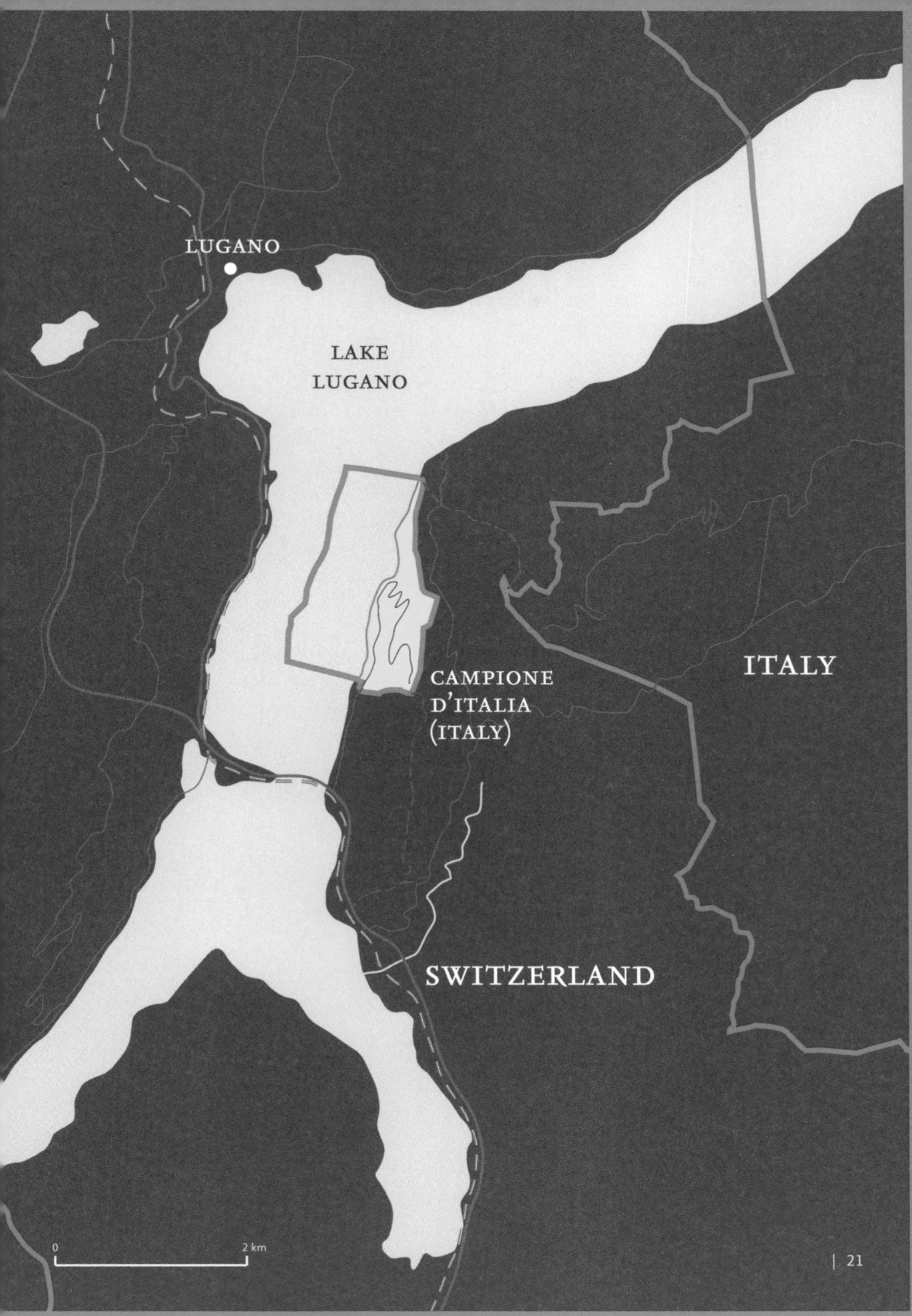

LUGANO

LAKE
LUGANO

CAMPIONE
D'ITALIA
(ITALY)

ITALY

SWITZERLAND

0

2 km

Lake Lugano lies in the far south of Switzerland, with the city that shares its name located on its northwestern shore. Two-thirds of the lake and its shores belong to Switzerland, and the remaining third to Italy. An important part of this Italian third is 'besieged' by the territory of Switzerland.

The part in question is the Italian town of **Campione d'Italia**, located on the eastern shore of Lake Lugano, southeast of the city of Lugano. The town covers about 1.6 square kilometres and is home to over 2,000 people. Although it is less than a kilometre away from the rest of Italy, high mountains prevent easy access to the home country. Therefore, Campione residents have to travel nearly fifteen kilometres to reach the nearest Italian town. The beautiful scenery that surrounds the glacial lake attracts many visitors to the area.

Owing to tourism and a large casino (the unique status of Campione allows free gambling and tax-free sales), this town is very rich. The casino is the largest single employer in the enclave. Founded in 1917 as a place for collecting information from foreign diplomats during the First World War, it is now the largest casino in Europe and provides sufficient income for Campione to prevent the need for tax. The casino is under state ownership, and operated by the local council.

Although Campione is a sovereign Italian territory, many of its services are more closely linked to Switzerland. For a start, the official currency is the Swiss franc, although the euro is usually accepted too. Switzerland is responsible for customs formalities, and cars use the licence plates of the Alpine Confederation. Likewise telephones: almost all phones go through Swiss operators, so that calls from the rest of Italy to Campione are treated as international. In terms of addresses, both Italian and Swiss postcodes are permitted. The residents of Campione have the right to use the services of Swiss hospitals, as if they were residents of Switzerland.

Historically, one of the most important moments for Campione occurred at the end of the eighteenth century, when the Canton of Ticino joined the Swiss Confederation. The residents of Campione decided to remain within Lombardy, which later became part of Italy. In the mid-1930s, by the decision of the then Italian Duce Benito Mussolini, the suffix d'Italia was added to the town's name, in order to further emphasise its allegiance to Italy.

During the Second World War, Campione largely became separated from the rest of Italy and functioned almost as a Swiss canton. As a result, this small town, unlike the rest of Italy, was not occupied during the war by the Germans, nor by the Allies after the war.

Some 200 kilometres northeast of Campione there is another Italian town tied to the Swiss system – **Livigno** (Livign in Lombard, Luwin in German). Although not an enclave, its limited travel connections with the rest of Italy mean that it has had the status of a 'duty-free' zone for some time.

VENNBAHN

BELGIUM | GERMANY

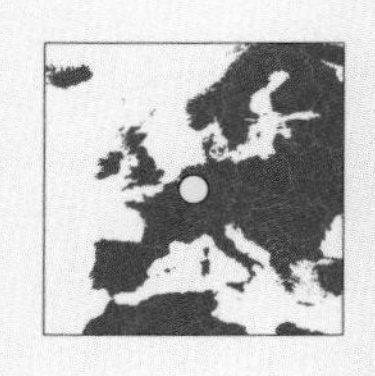

Enclaves that were created when the border railway's ownership passed from Germany to Belgium

50° 34' 56"N | 06° 13' 50"E

VENNBAHN
MÜNSTERBILDCHEN (GERMANY)
ROETGEN
ROETGENER WALD (GERMANY)
BELGIUM
RÜCKSCHLAG (GERMANY)
SIMMERATH
BELGIUM
BELGIUM
GERMANY
GERMANY
GERMANY
MÜTZENICH (GERMANY)
MONSCHAU
RUITZHOF (GERMANY)
VENNBAHN
0
3 km

Most of the world's enclaves were formed as a result of some historical 'gameplay': conquests, peace treaties, various border movements. Undoubtedly, among the strangest such territories, are the five within Belgium that belong to Germany – and all of these are separated from their home country by a single railway line, known as the **Vennbahn**.

At the end of the nineteenth century, what is now eastern Belgium was part of the German Empire. This area lies to the south of Aachen – the former capital of the Frankish emperor Charlemagne (Charles the Great) – and, until recently, it was traversed by a seemingly ordinary railway. Although the railway lay within Germany, its route took it right along the border with Belgium. And so it was until the end of the First World War.

Following the war, under the terms of the Treaty of Versailles, Belgium gained some territory from Germany. The railway passed through this land, with the exception of a few dozen kilometres, which remained in Germany. Belgium asked the Allies for the whole railway line, which was approved, and so the whole of the route came under Belgian sovereignty, along with a few metres either side of the track. In this way, some areas of German territory became isolated from the rest of the country, thus creating enclaves.

Over time, some of these enclaves have been absorbed into one or other country, some have changed shape, some have increased or decreased in size, and some have merged or separated. For a brief period there was even a small Belgian counter-enclave (an enclave within an enclave) within one of the German areas. Today, the enclaves that remain are Münsterbildchen, Roetgener Wald, Rückschlag, Mützenich and Ruitzhof. Because both Germany and Belgium are members of the European Union, the problem of defining the borders of them is considerably less of an issue than it was in the past. The extents of the remaining enclaves are very varied: the largest is about 1,200 hectares in area, while the smallest is only about 1.5 hectares. The populations vary considerably too: from 4 to about 2,500.

The railway was initially used for the transportation of coal and iron ore. Later it became a tourist attraction, but over time it lost that function as well. Today most of the railway has been dismantled and the larger sections of its route used as an increasingly popular cycle path and walking trail, passing through mostly flat green countryside.

At the beginning of the twenty-first century there were suggestions that, since it is no longer in use as a railway line, the land through which the Vennbahn passed should be returned to Germany. However, the Belgian and German prime ministers have recently pointed out that the border between Belgium and Germany has long been clearly defined and they see no reason to change that now.

Interestingly, the German language is predominantly spoken in the part of Belgium that surrounds these enclaves.

Vennbahn Trail, Belgium | Germany

BÜSINGEN AM HOCHRHEIN

GERMANY | SWITZERLAND

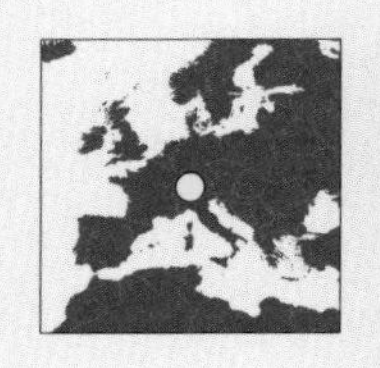

The German town that is not part of the European Union

47° 42' 22"N | 08° 41' 21"E

GERMANY

BÜSINGEN AM
HOCHRHEIN
(GERMANY)

SCHAFFHAUSEN

RHINE

SWITZERLAND

0 2 km

Büsingen am Hochrhein ('Buesingen on High Rhine') is a German town entirely surrounded by Swiss territory, namely the cantons of Schaffhausen, Thurgau and Zürich (the latter two being on the opposite side of the River Rhine). Büsingen occupies an area of about 7.5 square kilometres and has a population of about 1,500. It is separated from the rest of Germany by a strip of land, which is only 700 metres wide at its narrowest part.

Back in the seventeenth and eighteenth centuries, conflicts between Schaffhausen and Austria resulted in Büsingen becoming an Austrian territory surrounded by Schaffhausen. Germany took control of Büsingen in the middle of the nineteenth century, at the end of which Switzerland granted permission to the inhabitants of the town to sell their agricultural produce on its territory.

After the First World War, a referendum was held in Büsingen in which 96 per cent of the population expressed their desire to join Switzerland. Switzerland rejected this request as it lacked an appropriate territory to offer Germany in exchange. The population expressed its desire a few more times over the next two decades, but the outcome was always the same. During the Second World War, the Swiss police did not allow German soldiers to carry weapons within the territory of Büsingen, and after the war the town was temporarily occupied by the Allies (the French, to be precise), but only once Switzerland had approved.

The status of Büsingen has led to some unusual situations. For a start, the enclave is not part of the European Union, even though (West) Germany was one of the founding nations. This explains why Büsingen is part of the Swiss customs territory, together with Liechtenstein and the Italian enclave of Campione d'Italia. Although it is possible to make payments in euros, the most widely used currency is the Swiss franc, popular with the residents since most of them work in the surrounding Swiss towns and therefore receive their salaries in that currency.

Police protection for the enclave is provided by both Switzerland and Germany, with clearly defined numbers of officers from each country. There is a primary school for the lower years, after which the parents can decide whether their children should continue their education in Swiss or German schools. Postal and telecommunications services are provided by both Swiss and German companies, with postal and telephone area codes from both countries.

Although Büsingen territorially belongs to the German city of Konstanz, where vehicle licence plates are denoted with KN, Büsingen has its own licence plates with the marking BÜS. This allows for simpler control by the Swiss customs, as these cars are treated as local. In fact, however, most residents of Büsingen have their cars registered in the Swiss canton of Schaffhausen, meaning that licence plates with the letters BÜS are among the rarest in Germany and can be found only on a few hundred vehicles.

And finally, a sporting oddity: the local football club is the only German club competing in the Swiss league.

Café in Büsingen am Hochrhein, Germany | Switzerland

BAARLE

BELGIUM | NETHERLANDS

The 'citizenship' of a house is determined by the position of its front door

51° 25' 59"N | 04° 55' 01"E

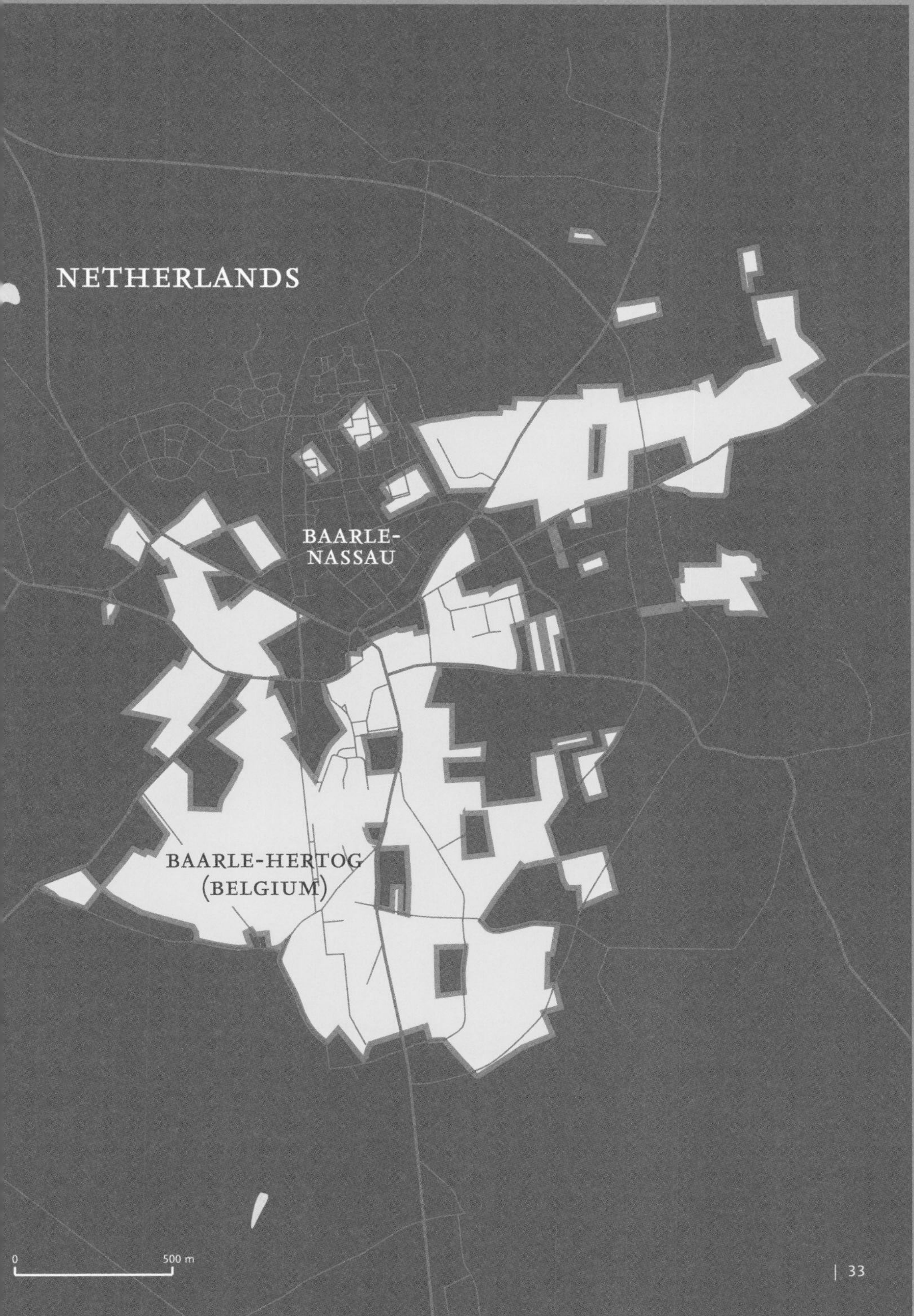

NETHERLANDS
BAARLE-
NASSAU
BAARLE-HERTOG
(BELGIUM)
0
500 m

Baarle is a small town in the south of the Netherlands, close to the border with Belgium. Although the international boundary runs to the south of the town, it also passes through the town itself.

The town consists of two parts: the Dutch **Baarle-Nassau** and the Belgian **Baarle-Hertog.** The Belgian part includes more than twenty enclaves within the Dutch section of the town, while inside these Belgian enclaves there are approximately ten Dutch counter-enclaves. This means that the border intersects some of the streets several times, while some houses are partly in Belgium and partly in the Netherlands.

The complexity of the border is the result of a series of divisions, contracts and exchanges of territory, many of which have their roots in the Middle Ages. Although the divisions were largely confirmed by the mid-nineteenth century, the border was not completely defined until 1995, when the extent of the smallest Belgian enclave – an uninhabited piece of agricultural land with an area of about 2,600 square metres – was finally established.

In order to make the situation a little clearer (primarily because of the large number of tourists who visit), boundaries are often physically marked on the streets themselves. Also, as Dutch and Belgian house numbering differs, there is sometimes a Dutch or Belgian flag alongside the house number. In cases where the boundary cuts through a house, its 'citizenship' is determined on the basis of whose territory the front door is facing. Throughout history – and accompanying numerous changes of tax rates in the Netherlands and Belgium – relocating the front door was not unusual, in order that the owner could pay less tax.

The Belgian enclaves range in size from 0.2 hectares to about 153 hectares, while the Dutch enclaves from about 0.28 hectares up to a little more than five hectares. In addition to the enclaves within the city itself, the Belgian Baarle-Hertog also has several small enclaves around it.

The borders of these enclaves are fully open, but anybody crossing them should be aware of the things that are permitted in one country, but forbidden in another. For example, in the Belgian Baarle there are a number of fireworks shops, but the free sale of fireworks in the Netherlands is prohibited. This fact is often taken advantage of by the Dutch on the eve of their national holidays, when many of them cross the border to buy fireworks for the celebrations. Another interesting practice was relatively common in previous years: because the closing time for restaurants was earlier in the Netherlands than in Belgium, at those establishments through which the border passes, diners would simply transfer to the Belgian half when the Dutch closing time came, and continue their socializing there.

Also, it is interesting to note that during the First World War, during which the Netherlands was neutral and Belgium was occupied by Germany, Baarle-Hertog was the only free part of Belgium.

JUNGHOLZ

AUSTRIA | GERMANY

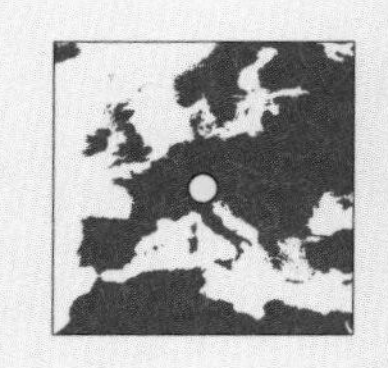

Linked to its home country at a mountain-top quadripoint

47° 34' 42"N | 10° 27' 20"E

GERMANY

JUNGHOLZ
(AUSTRIA)

MT SORGSCHROFEN

AUSTRIA

0 1 km

Jungholz is an Austrian village in the state of Tyrol. It lies just over 1,000 metres above sea level, covers an area of approximately seven square kilometres and has around 300 inhabitants.

The village is almost entirely surrounded by German territory, but what makes it unusual is that it is connected to its home country at a single point: the summit of Mt Sorgschrofen, at a height of over 1,600 metres.

The border was defined back in the mid-nineteenth century by a treaty between the Kingdom of Bavaria and the Austrian Empire, although it had existed in a similar form as early as the fourteenth century. After the establishment of the border, Jungholz was economically linked to Bavaria, and later to Germany. This meant that border control was carried out through German customs authorities, while both Austrian and German postal codes and telephone area codes were used equally. At the beginning of the twenty-first century, Austria wanted to abolish the German postal codes in Jungholz, but protests by the residents forced a back-down. Conversely, there are plans to abolish the German telephone area code, leaving only the Austrian one in use. Until the introduction of the euro, as a common European currency, only the German Deutschmark was used, and after the establishment of the eurozone, German tax laws still applied. Interestingly, because it has such a specific status, there are branch offices of three of the largest Austrian banks in Jungholz, making it a place with the highest number of banks per capita.

There is another curiosity related to Jungholz: the existence of one of the rare world quadripoints, namely a point where four borders meet. In this case it is two Austrian borders meeting two German borders, a junction sometimes called a **binational quadripoint** or a **boundary cross**. There are only two other such quadripoints in the world: in the Belgian–Dutch Baarle and in one recently abolished Indian–Bangladeshi enclave.

As far as economic ties with Germany are concerned, a similar situation, until recently, existed in three villages in the nearby **Kleinwalsertal**, a valley in Austria's westernmost state of Vorarlberg. Due to the nature of the terrain, this valley is almost completely cut off from the rest of Austria, since the only way out of it is through Germany. For this reason, Kleinwalsertal was granted special economic status in the first half of the nineteenth century (duty-free zone, German customs control, use of the Deutschmark), which it kept until the beginning of the twenty-first century when Austria entered the European Union, signed the Schengen Agreement and accepted the euro.

CYPRUS
AND ITS BORDERS

The playground of the rich and famous that became a ghost town

35° 09' 01"N | 33° 29' 23"E

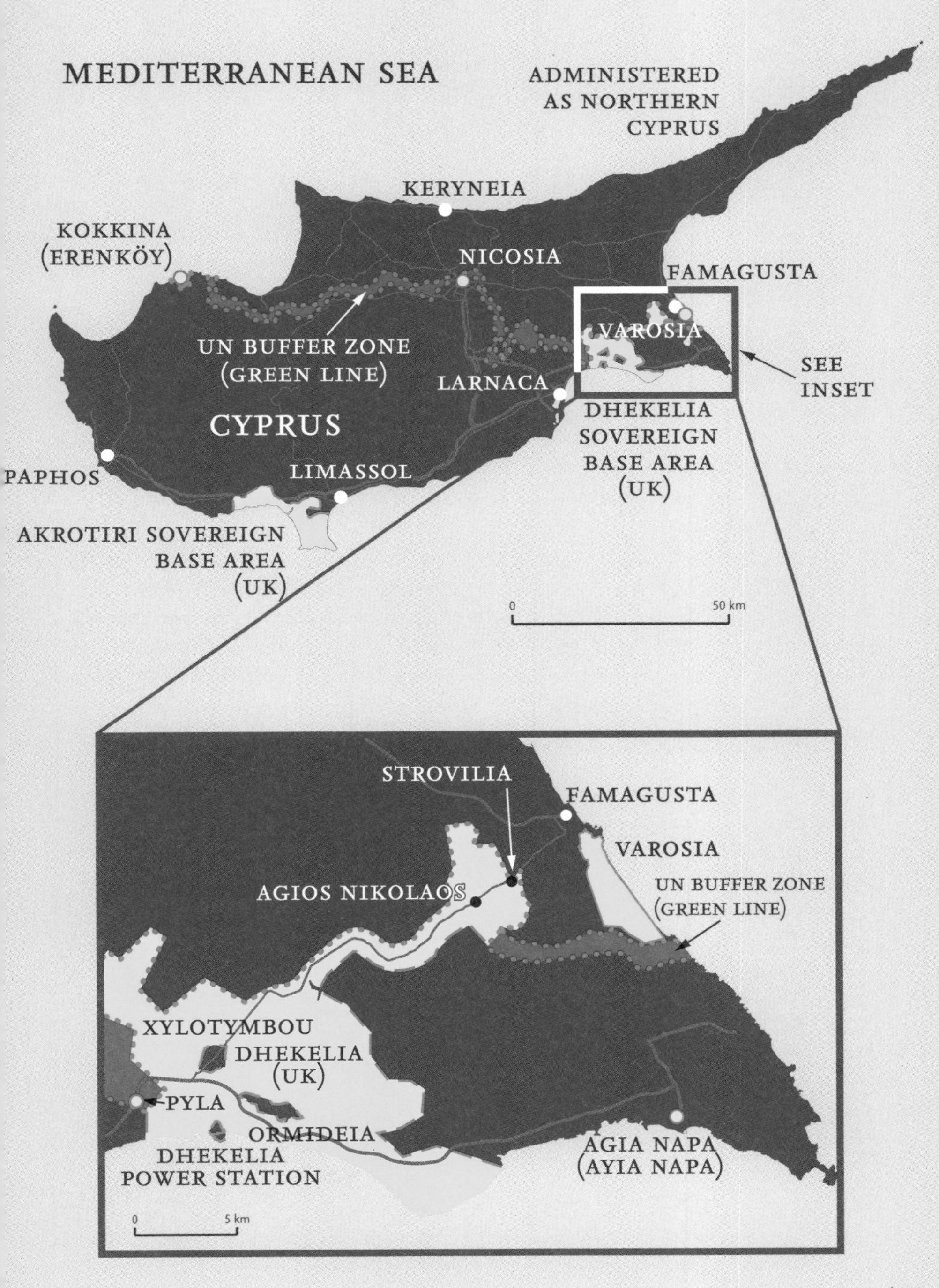
MEDITERRANEAN SEA
ADMINISTERED AS NORTHERN CYPRUS
KERYNEIA
KOKKINA (ERENKÖY)
NICOSIA
FAMAGUSTA
VAROSIA
UN BUFFER ZONE (GREEN LINE)
LARNACA
SEE INSET
CYPRUS
DHEKELIA SOVEREIGN BASE AREA (UK)
PAPHOS
LIMASSOL
AKROTIRI SOVEREIGN BASE AREA (UK)
0
50 km
STROVILIA
FAMAGUSTA
VAROSIA
AGIOS NIKOLAOS
UN BUFFER ZONE (GREEN LINE)
XYLOTYMBOU
DHEKELIA (UK)
PYLA
ORMIDEIA
DHEKELIA POWER STATION
AGIA NAPA (AYIA NAPA)
0
5 km

A quick glimpse at a map of the eastern Mediterranean may give the impression that Cyprus is an island country with no unusual political-geographical characteristics. However, such a glimpse would be misleading, since the island, although only covering about 9,250 square kilometres, is the home of no less than four political territories and numerous borders.

For a long time, the history of Cyprus has been affected by the relationship between Greece and Turkey, but it was the gaining of independence from the United Kingdom in 1960 that laid the foundations for the first of Cyprus's numerous internal borders to be created. Two large military bases, covering a total of 3 per cent of the area of Cyprus, remained under the governance of the UK, as a **British Overseas Territory**.

Over the next fifteen years or so, the situation in Cyprus was rather unstable, with frequent calls from Greece to have the island united with Greece, coupled with occasional threats from the northern neighbour, Turkey, to use military intervention to protect the Turkish population there. Eventually, in 1974, this threat was realized, and after an invasion by the Turkish army, the northern third of the island was quickly occupied. This event also triggered the gradual formation of the next border, the one between the area administered as northern Cyprus by Turkey and the rest of the island still governed by Greece.

The United Nations became aware that without the establishment of a neutral zone, there would be no peace in Cyprus. And so the **United Nations Buffer Zone** was established along the entire length of the line dividing the Greek and Turkish forces. This area, also known as the **Green Line**, is about 180 kilometres long, with a width varying from 7.5 kilometres at its broadest part to just a few metres in some parts of Nicosia, the capital city. This UN zone represents the fourth political entity of the divided island, with its own laws and border controls. About 10,000 people live and work within the area, which covers almost 350 square kilometres. As well as the inhabited villages, there are also deserted settlements and numerous hastily abandoned facilities close to the dividing line in Nicosia, such as car showrooms still displaying 'new' cars that date back to the 1970s. The movement of civilians who are living and/or working in the Green Line is permitted across most of the zone, though there are certain parts that are only accessible to the 'Blue Helmets'. Although the UN recognizes the sovereignty of the Republic of Cyprus over the Green Line, it is essentially a separate political and territorial element of this divided island.

And as if the different political entities, with their borders, are not enough for a relatively small island nation, several other enclaves, semi-enclaves and corridors are also found there.

The first unusual enclave is the town of **Kokkina** (the Greek name) or **Erenköy** (the Turkish name). This small former Turkish town is presently completely abandoned by the civilian population, and only a small Turkish garrison is found there. It is situated in the western part of Cyprus, surrounded by the UN Green Line and the Greek territory, at a distance of seven kilometres from the rest of the area administered as northern Cyprus.

Two British military bases, which together form the **Sovereign Base Areas of Akrotiri and Dhekelia**, constitute a British Overseas Territory, under the governance of the administrator, who at the same time is the Commander of the Armed Forces. These bases remain a part of the United Kingdom, but, according to the treaty with Cyprus, they are allowed to be used only for military, not commercial, purposes. Currently, this territory is the only part of the UK where the euro, and not the pound, is used. Akrotiri, or the Western Sovereign Base Area (WSBA), is situated in the south of Cyprus, close to Limassol. Dhekelia, officially known as the Eastern Sovereign Base Area (ESBA), is situated in the southeast of the island, near Larnaca. The majority of this base is located on the coast, but there is a road connecting it to a communication base in the former village of Agios Nikolaos. This road, a sort of corridor of some ten kilometres in length and only a hundred or so metres wide, also acts as a separation zone between the Greek and Turkish forces.

Within the main part of Dhekelia there are officially three Greek-Cypriot enclaves, although in reality there are four. Two large enclaves are the villages of **Ormideia** (179 hectares, 5,000 inhabitants) and **Xylotymbou** (95 hectares, 3,600 inhabitants). These two villages, tucked comfortably within the British base, represented a peaceful oasis during the Turkish–Greek conflicts in Cyprus, and it was there that many Greek refugees from the northern part of the island found their safe haven. The third enclave within the Dhekelia base covers the area of the thermal power plant of the same name. The power plant itself is situated on the coast, with a settlement to the north where the employees and refugees from northern Cyprus live. As the power plant and the settlement are divided by a road under the sovereignty of Britain, they effectively form two separate Greek-Cypriot enclaves. As the power plant is situated at the coast, with no territorial waters appertaining to it, it is therefore completely surrounded by the land and sea of the British base.

In a broader sense, even the well-known tourist resort of **Agia Napa (Ayia Napa)** and its surroundings could be regarded as a semi-enclave, since it is separated from the remaining Greek-Cypriot territory by the British base Dhekelia, although free access is possible via the sea.

Apart from these enclaves, there are several other places of interest in Cyprus.

The village of **Pyla** is situated within the UN Green Line and represents a rare example of co-habitation of the Greek and Turkish communities in Cyprus.

In the east of Cyprus, on the big bay of Famagusta, lies the town of **Varosia** (more precisely, a southern suburb of Famagusta). Until 1974, and the advance of the Turkish forces as far as Famagusta, Varosia was the best-known and most exclusive resort in Cyprus, a sort of Cypriot Monaco. The biggest stars of the time, such as Elizabeth Taylor, Richard Burton, Raquel Welch and Brigitte Bardot, were regular guests there. Following the fierce battles in Famagusta, the Greek population fled to the southern part of Cyprus, with the hope that they could return to their homes once the conflict had ended. However, that did not happen:

the Turkish authorities fully enclosed Varosia and proclaimed it a restricted zone. This status has remained to the present day, and Varosia has become a ghost town. Its buildings are slowly collapsing, and greenery has taken over the streets of the former jet-set centre of Cyprus.

Close to Varosia is another unusual settlement. At the border between the Dhekelia military base and northern Cyprus, lies the tiny hamlet of **Strovilia** (**Akyar** in Turkish), being the size of a football field and with approximately twenty-five inhabitants. The peculiarity of this village is that, during the conflict in the 1970s, the Turkish forces thought that it was within the British base, so they did not occupy it. When, after several weeks, they realized their mistake, the UN forces prevented any further attacks. And so, for the next twenty-five years, Strovilia occupied the unusual position of being the only part of southern Cyprus that directly, with no interlying United Nations 'Green Line' buffer zone, bordered the area administered as northern Cyprus. The situation changed at the beginning of the twenty-first century when the Turkish forces, despite opposition from Cyprus, the UK and the UN, managed to occupy a small part of the village. In response to this act by Turkish forces, the Greeks introduced the blockade of Kokkina. At the beginning of 2019, the Turkish forces brought pressure on Strovilia again, by erecting fences and blockades around the village, and by informing the population that they were now living in the area administered as northern Cyprus.

The city of **Nicosia** (Lefkosia in Greek, Lefkoşa in Turkish) is also worthy of mention. It is the capital of Cyprus, but it is also the city through which the Green Line passes, and where the UN holds governance. Finally, Cyprus also had a time border during 2016, when the area administered as northern Cyprus decided not to switch to summer daylight saving time, meaning the island had two different time zones for a while. The use of daylight saving time in the northern part of Cyprus was resumed in 2017, upon the request of the local population.

EXCLAVES OF RUSSIA

SAN'KOVO-MEDVEZH'YE
LUTEPÄÄ TRIANGLE
SAATSE BOOT
KALININGRAD

There is a road on which vehicles are not allowed to stop

San'kovo-Medvezh'ye 52° 28' 45"N | 31° 33' 42"E
Lutepää Triangle 57° 55' 26"N | 27° 40' 56"E
Saatse Boot 57° 54' 14"N | 27° 42' 15"E
Kaliningrad 54° 43' 05"N | 21° 38' 18"E

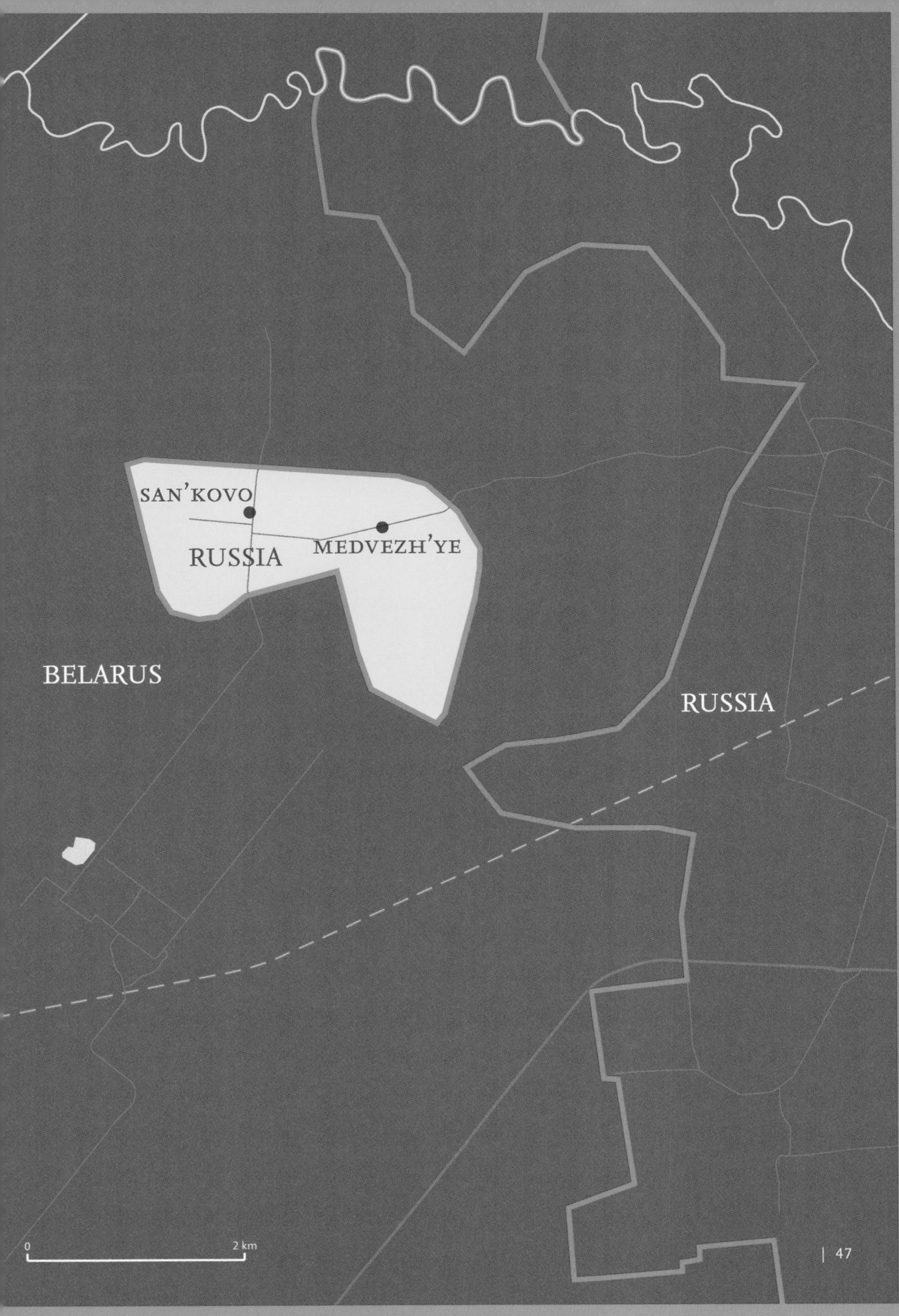

SAN'KOVO
MEDVEZH'YE
RUSSIA
BELARUS
RUSSIA
0
2 km

Russia is the largest country in the world and extends into two continents, eastern Europe and northern Asia. It consists of many autonomous republics, regions and other territories. However, even such a large country has a few parts of its domain that are separated from the homeland.

One such piece of land is the **San'kovo-Medvezh'ye** enclave (Саньково-Медвежье in Russian), within the territory of Belarus (see map on p47). It lies approximately thirty-five kilometres east of the city of Homyel'/Gomel', and about 530 kilometres southwest of Moscow. The enclave occupies an area of roughly four and a half square kilometres, and the population is zero. Although it contains two small villages, they are abandoned. These villages were evacuated after an explosion at the distant Chernobyl Nuclear Power Plant, 150 kilometres away; housing and any other economic activity in the area are now completely forbidden.

There are several different theories as to how this enclave was formed, but it is a fact that it is now rarely shown on maps of Russia, partly because it is too small for the scales used to show the rest of the country, and partly because of a lack of interest by the Russian authorities in this area. Nowadays, this village is mainly only visited by thieves – almost all of the houses have been largely 'disassembled', with doors, windows, and even bricks and pipes having long been stripped out – and poachers, who use this enclave for hunting outside of the hunting season in Belarus. Since the Russian police do not go there, and the Belarus police do not have jurisdiction over the area, this situation will remain until the two Slavic neighbours agree on the future of the enclave.

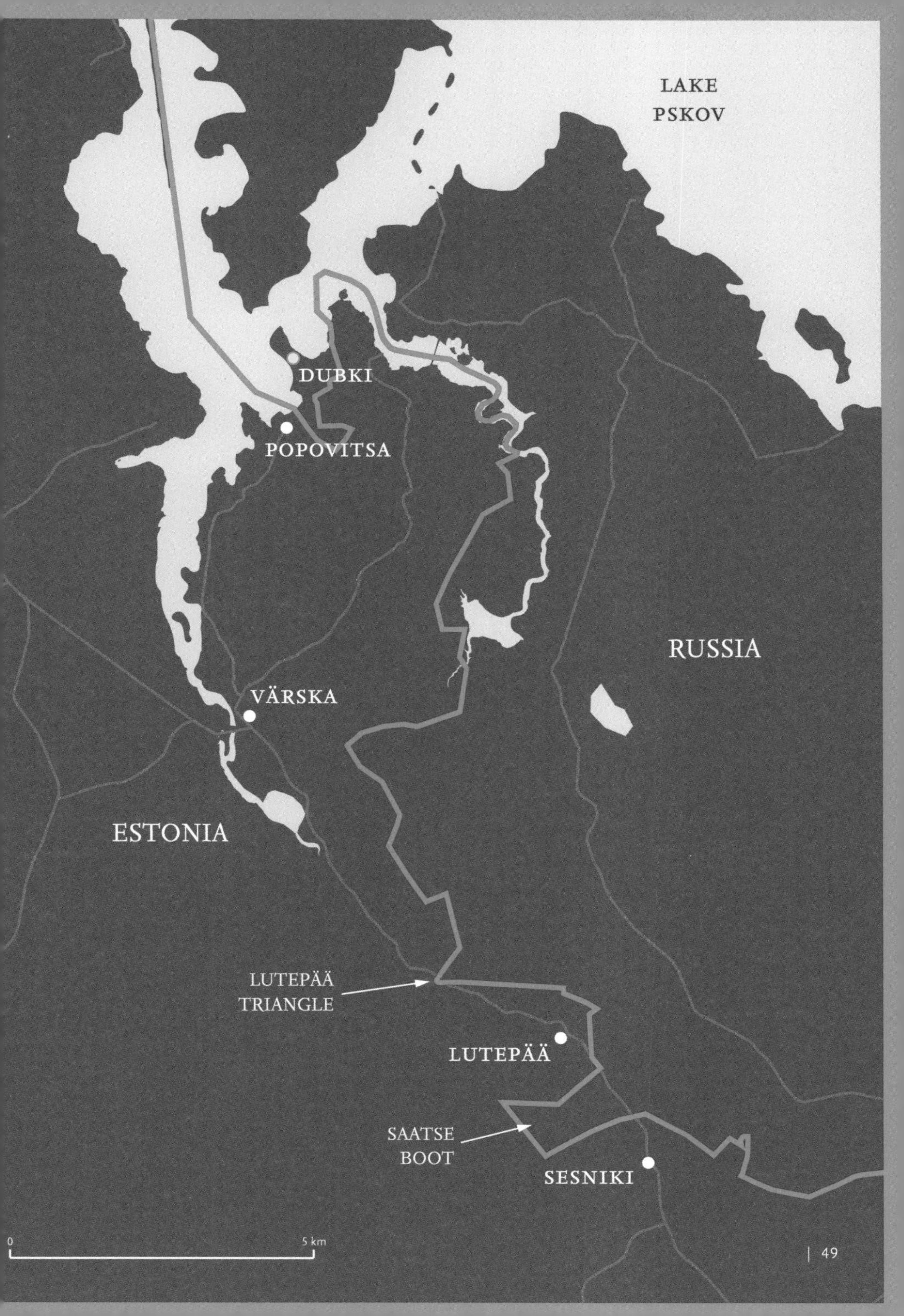
LAKE
PSKOV
DUBKI
POPOVITSA
RUSSIA
VÄRSKA
ESTONIA
LUTEPÄÄ
TRIANGLE
LUTEPÄÄ
SAATSE
BOOT
SESNIKI
0
5 km

On the shores of Lake Pskov, the southern part of Lake Peipus, lying on the border of Estonia and Russia, there is a tiny village with only ten or so inhabitants – **Dubki** (or Tupka, according to the language of the local Seto people, related to the Estonians) (see map on p49). The village is located on a peninsula that was part of independent Estonia from the 1920s until the end of the Second World War. After the war it was annexed to Russia along with a few surrounding territories. Today, this peninsula is accessed by land from Russia only through Estonia. However, it is connected to the home country by waterway, via the lake, which means that the village of Dubki is not a classic enclave, though it is certainly a Russian exclave surrounded by the territory of Estonia.

The road that leads from the Dubki peninsula southwards is also of interest. This Estonian road starts from the village of Popovitsa, on the border with Dubki enclave, then goes south through the villages of Värska, Lutepää and Sesniki. What makes this local road unusual is that it passes through the territory of Russia at two places:

- Between the villages of Värska and Lutepää, through about fifty metres of the Russian territory known as the '**Lutepää Triangle**' (Lutepää kolmnurk in Estonian)
- Between the villages of Lutepää and Sesniki, for about one kilometre, the road passes through the '**Saatse Boot**' (Saatse saabas in Estonian; Саатсеский сапог in Russian).

Although no special permission or visa is necessary to drive through these parts of Russia, vehicles are not allowed to stop. Even if they run out of petrol, the driver will be questioned by the police. It is not permitted to walk through this part. The fact that these Estonian villages are connected by a road that passes through a neighbouring country effectively makes them enclaves. In 2008 a new road was built that goes around the triangle and boot, but it is twenty kilometres longer than the 'Russian road'. Recent negotiations between Estonia and Russia have made it likely that these awkward borders will be 'straightened' by an exchange of territories.

LITHUANIA

BALTIC
SEA

KALININGRAD

RUSSIA

POLAND

0 25 km

The biggest enclave in Europe, the **Kaliningrad** region (see map on p51), has an area of about 15,000 square kilometres and almost a million inhabitants. It is surrounded by Lithuania to the north and east, Poland to the south and the Baltic Sea to the west. The closest point to Russia is more than 350 kilometres by air. Up until the Second World War, Kaliningrad was a significant Prussian city known as Königsberg. After the war, Russia gained the northern half of German East Prussia, while Poland got the southern half. The Germans, who were practically the only inhabitants of this region until then, were expelled and the territory populated with Russians and some Ukrainians and Belarusians. The city's name was changed to Kaliningrad in memory of Mikhail Kalinin, who was leader of Russia and the Soviet Union from 1919 until his death in 1946. Kaliningrad's economy benefits from its ice-free ports and its proximity to the European Union. It also has 90 per cent of the world's amber reserves and there are advances being made in industry (for example a few significant car manufacturing plants) and tourism.

The **Crimean Peninsula**, recently annexed by Russia, may also be considered to be an enclave, because it does not have a land connection with Russia, although the eighteen-kilometre-long Crimean Bridge was constructed by Russia to span the Strait of Kerch between Russia and Crimea.

Far to the south, in Azerbaijan, there are (or were) two villages that were Russian enclaves in this Caucasian country. According to some sources, in the mid-1950s Azerbaijan leased the village of **Khrakhoba** to Russia, but over time its inhabitants moved to the nearby Russian Republic of Dagestan. There is another former Russian enclave, the village of **Uryanoba**, whose inhabitants have mostly left their homes. The majority of the populations of both villages are Lezgins, one of the many Caucasian ethnic groups. The Russian president Medvedev officially signed an Act for the return of these two villages to Azerbaijan.

Another interesting area, within the territory of Kazakhstan, could be considered a (temporary) exclave of Russia. It is an elliptical-shaped piece of land surrounding the **Baikonur Cosmodrome**, which Kazakhstan has leased to Russia until 2050. While this agreement is in force, Baikonur has the status of a 'city of federal importance', the same as Moscow, St Petersburg and Sevastopol' in Crimea (Ukraine does not recognize Russia's sovereignty over Crimea and Sevastopol'). This 'cosmic' exclave covers an area of about 6,000 square kilometres, and the mayor is appointed by mutual consent of the presidents of Kazakhstan and Russia, upon proposal by Russia.

ENCLAVES IN CENTRAL ASIA

100,000 people were separated from their home countries by the dissolution of a nation

40° 09' 04"N | 71° 17' 47"E

AZAKHSTAN
KYRGYZSTAN
UZBEKISTAN
SHKENT
SARVAN (TAJ.)
BARAK (KYRG.)
OSH
CHON KARA (UZBEK.)
DZHANGAIL (UZBEK.)
SHAKHIMARDAN (UZBEK.)
KAYRAGACH (TAJ.)
SOKH (UZBEK.)
VORUKH (TAJ.)
KYRGYZSTAN
TAJIKISTAN
0
50 km
AFGHANISTAN

Within many countries, there are 'internal' enclaves, parts of individual provinces, states or federal states within other such areas. As long as a country is united, these internal enclaves do not represent a (significant) problem. However, if such a country breaks up, the internal enclaves become true enclaves, with all the difficulties this creates.

This was the case with the break-up of the USSR and the proclamations of independence of its Central Asian republics. During the Soviet Union, the borders of its republics were defined by the leaders of the Communist Party, who decided that the borders in Central Asia should be determined, to a large extent, along linguistic lines. So, in theory, a village in which language X is spoken should belong to republic X, even if it is surrounded by republic Y. During the USSR this did not present a problem, but today, there are several enclaves within the independent Central Asian countries of Uzbekistan, Tajikistan and Kyrgyzstan. Between them, they are inhabited by almost 100,000 people.

Most of the enclaves are in Kyrgyzstan:

- The westernmost enclave, **Kayragach** (Western Qal'acha), with an area of less than one square kilometre, belongs to Tajikistan. It is uncertain whether this enclave is inhabited at all.
- Another enclave of Tajikistan in Kyrgyzstan is **Vorukh**. This is a relatively large and populated enclave – almost 100 square kilometres and seventeen villages with a total of over 25,000 inhabitants, of whom 95 per cent are Tajik.
- The largest Uzbekistan enclave is **Sokh** or So'x, with an area of over 230 square kilometres, a width varying between three and thirteen kilometres and a length of thirty-five kilometres. One of Kyrgyzstan's main motorways passes through this enclave, whose 43,000 (or, according to some sources, over 70,000) inhabitants are almost exclusively Tajik. It would appear that, of all the Central Asian enclaves, this is the only one with some sort of local parliament. The other enclaves are organized as ordinary villages, without separate state bodies.

- Another relatively large Uzbek enclave is **Shakhimardan**, located approximately twenty kilometres south of the Uzbek border. A population of 6,000, of which over 90 per cent are Uzbek, lives in this territory of about forty square kilometres.
- **Chon-Kara** or Chon-Qora is a small Uzbek enclave, just north of Sokh. It consists of two Uzbek villages, with an area of about three square kilometres, and is about three kilometres from the Uzbekistan border.
- **Dzangail** is a small Uzbek enclave, about one kilometre long and a kilometre from the Uzbek border. It is unclear whether this enclave still exists, even though it is still marked on some maps.

Uzbekistan is home to the two remaining enclaves:

- **Sarvan** is located about 100 kilometres east of Tashkent, the capital of Uzbekistan. This Tajikistan enclave is about one and a half kilometres from its home country, with an area of eight square kilometres and less than 500 inhabitants. Although Tajikistan claims that it governs this enclave, this does not seem to be true – according to various sources, it appears that all the power has been in the hands of Uzbekistan since at least the beginning of this century, so it is debatable whether this still exists as an enclave at all.
- **Barak** is a Kyrgyzstan enclave in Uzbekistan ... or maybe not? Another confusing enclave in Central Asia: according to some sources, this small enclave is located a few kilometres north of the Kyrgyzstan–Uzbek border, near the city of Osh. Other sources claim that it is not an enclave (or at least not any more), but a Kyrgyzstan border village.

Obviously, the situation with enclaves in Central Asia (or, more precisely, in the Fergana valley) is very confusing. The three countries are attempting to finally define their borders, which may result in the elimination of some of these enclaves or their merging with their home countries. Or, one never knows – perhaps the talks will result in some new enclaves and borders?

MADHA AND NAHWA

OMAN | UNITED ARAB EMIRATES

The nationalities of these enclaves were chosen by their inhabitants

25° 16' 47"N | 56° 16' 50"E

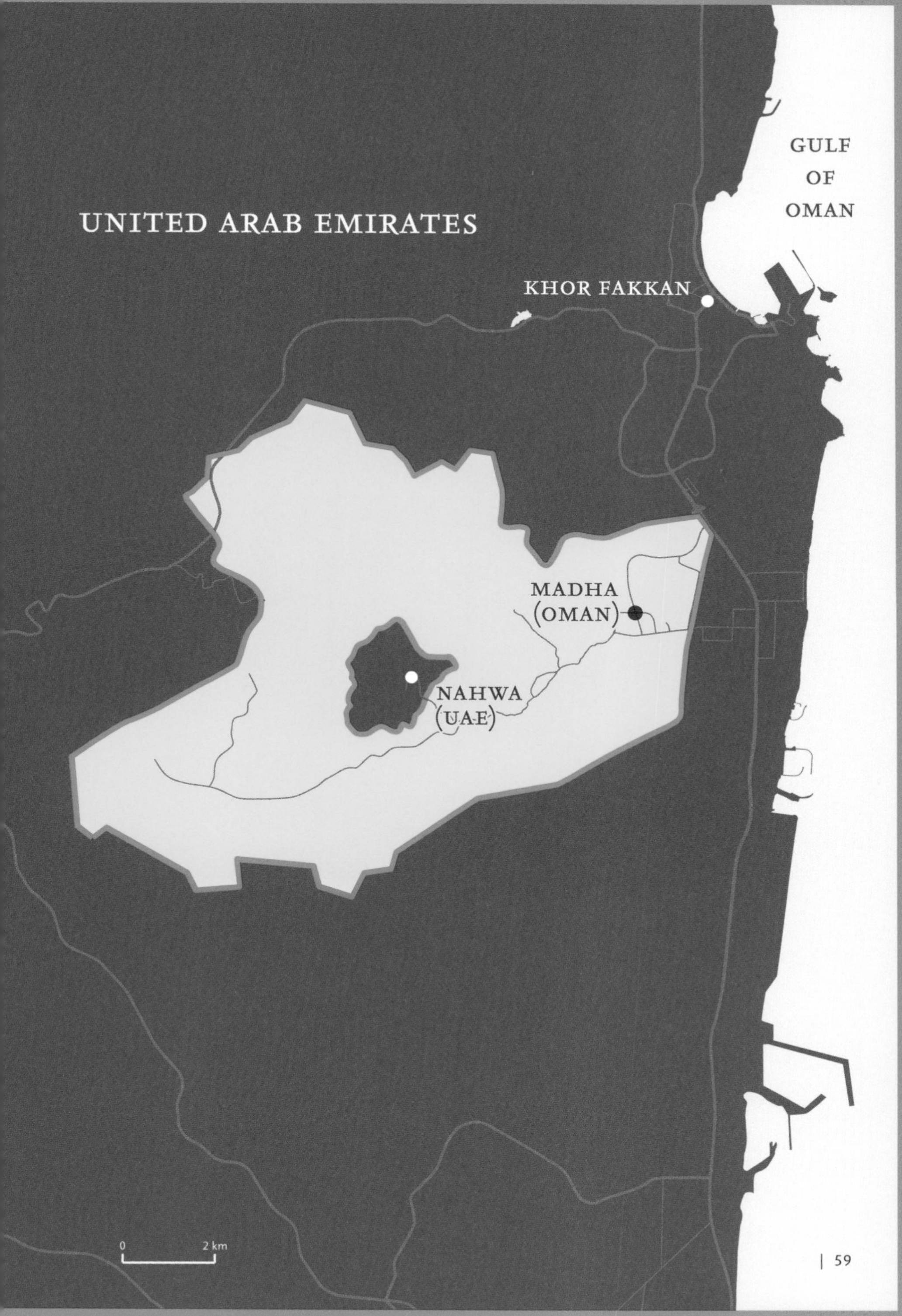

UNITED ARAB EMIRATES

GULF
OF
OMAN

KHOR FAKKAN

MADHA
(OMAN)

NAHWA
(UAE)

0 2 km

The Arabian Peninsula is a vast peninsula in Southwest Asia, rich in oil and gas. Since desert covers the greater part of this peninsula, the borders between the countries within it have, to a large extent, yet to be clearly defined. This is particularly the case with the borders between Saudi Arabia and its southern and southeastern neighbours: Yemen, Oman and the United Arab Emirates (UAE).

Even within the borders on the Arabian Peninsula that are relatively clearly defined, there is still an oddity in one of them: the Oman enclave within the UAE contains the small town of **Madha**, which in turn contains the UAE counter-enclave of the village of **Nahwa**.

Madha, covering seventy-five square kilometres and with about 3,000 inhabitants, is quite a large enclave. Most of it is uninhabited, except for several small, scattered villages and the small town of the same name, also known as New Madha, on the eastern side of the enclave. This small town possesses all that is necessary for regular life, including a police station, a school, banks, power and water supply, and even an airport.

Less than ten kilometres west of Madha, along a good but winding road, lies the UAE counter-enclave of Nahwa. This enclave consists of two small towns: the very rich New Nahwa, with about forty houses, a police station, its own clinic and ordered streets; and the poor, haphazard and unpaved streets of Old Nahwa.

The unusual situation with this enclave and counter-enclave evolved from a democratic decision made by the local tribes during the period between the two world wars, when some tribes opted to become annexed to Oman, and some to the UAE. The acceptance of their 'referendum' decision resulted in these borders.

North of these two territories, there is another small Omani pene-enclave (semi-enclave), separated from the home country by the land of the UAE and the waters of Oman Bay. It is the extremely rugged coastline of the **Musandam Peninsula**, which strategically controls the Strait of Hormuz (through which one third of the world's oil tankers pass). This area is home to the Kumzari language, the only language from the Iranian group that is spoken on the Arabian Peninsula. The high mountains and magnificent fjords of the Musandam Peninsula represent a contrast to the mostly flat, sweeping coastline of the Arabian Peninsula.

ARMENIA–AZERBAIJAN BORDER

The autonomous region fighting for independence

39° 31' 34"N | 45° 31' 0"E

TBILISI

GEORGIA

ASKIPARA
(AZERBAIJAN)

BARXUDARLI
(AZERBAIJAN)

ARTSVASHEN
(ARMENIA)

KIROVABAD

AZERBAIJAN

YEREVAN

ARMENIA

KARKI
(AZERBAIJAN)

RKEY

NAGORNO-KARABAKH
(DAĞLIQ QARABAĞ)

NAKHCHIVAN
(AZERBAIJAN)

IRAN

0 50 km

Complicated borders and an unwillingness to compromise create excellent sources for misunderstandings, conflicts and even wars. Such a situation occurred on the southern slopes of the Caucasus Mountains following the dissolution of the USSR. Centuries of tension between Christian Armenia and Muslim Azerbaijan were exacerbated by the establishment of the autonomous provinces and enclaves during the time of the Soviet Union, and even more so with the declaration of independence of the former Soviet republics.

As the former republics became independent during the dissolution of the USSR, so some of the autonomous provinces also sought the freedom to decide their own futures. Such was the case with **Nagorno-Karabakh**, an autonomous region within Azerbaijan populated almost exclusively by Armenians. Following the independence of Armenia and Azerbaijan in 1991, Nagorno-Karabakh declared its independence as the Republic of Artsakh. This led to a bloody war over the region between the newly declared Republic, supported by Armenia, and Azerbaijan. This conflict ended with the truce of 1994, with the Republic of Artsakh in control of the former autonomous region and also much of the surrounding area up to the Armenian and Iranian borders. The truce, with occasional interruptions, lasted until late 2020, when war erupted again. This war came to an end following a Russian-brokered deal, under which Azerbaijan regained control of much of the Republic of Artsakh, including part of the former autonomous region itself.

The situation in the Soviet republics of Armenia and Azerbaijan was already complicated enough: within its borders, Azerbaijan had the autonomous Armenian region of Nagorno-Karabakh and one Armenian enclave, as well as its own exclave. On the other hand, Armenia had three Azerbaijani enclaves.

Artsvashen (the 'Eagle City') was an Armenian small town-enclave in the northwest of Azerbaijan. It covered an area of about forty square kilometres. Although this town, 'according to the documents', still belongs to Armenia, it has been exclusively populated by Azerbaijani refugees from other regions since the Armenian–Azerbaijani war in the early 1990s.

Azerbaijan has another large exclave, south of Armenia and north of Iran, named **Nakhchivan**. After the First World War and the fall of the Ottoman Empire, the USSR occupied the entire area south of the Caucasus Mountains. Nakhchivan became part of Azerbaijan, although it was populated mainly by Armenians until the nineteenth century and the genocide over this nation. Today, Nakhchivan has the status of an autonomous republic within Azerbaijan, and still suffers the consequences of conflict and closed borders between Armenia and Azerbaijan. As Nakhchivan borders with Armenia, Iran and Turkey, it is not a true enclave.

It is interesting to note that Nakhchivan, the Azerbaijani exclave, has its own exclave within Armenia. Until the war of the 1990s, **Karki** was an enclave covering an area of about twenty square kilometres and populated by Azerbaijanis. Nowadays, also known as Tigranashen, it is a village populated by Armenians, many of whom are refugees from Azerbaijan. The majority of the former Karki inhabitants moved to Nakhchivan, where they established a new village, New Karki.

Barxudarli is a former Azerbaijani enclave in the north of Armenia. Currently, this enclave covers an area of over twenty square kilometres. It is populated exclusively by Armenians, whereas the Azerbaijani, partly by their own will, but mostly under pressure, left the enclave and now live in the home country. Not far from Barhudarli there were another two miniature unpopulated Armenian enclaves.

Still further to the north, there is another former Azerbaijani enclave, **Askipara** (Əskipara). This village was almost completely devastated during the war, and only the foundations of destroyed houses remain.

TOMB OF SULEYMAN SHAH

SYRIA | TURKEY

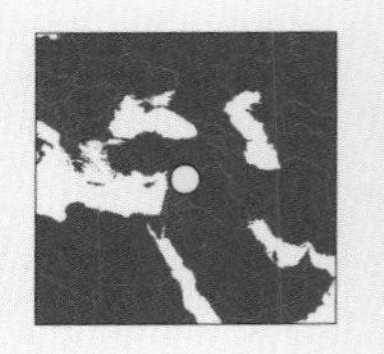

The interment that has yet to find peace

35° 57' 03"N | 38° 23' 08"E

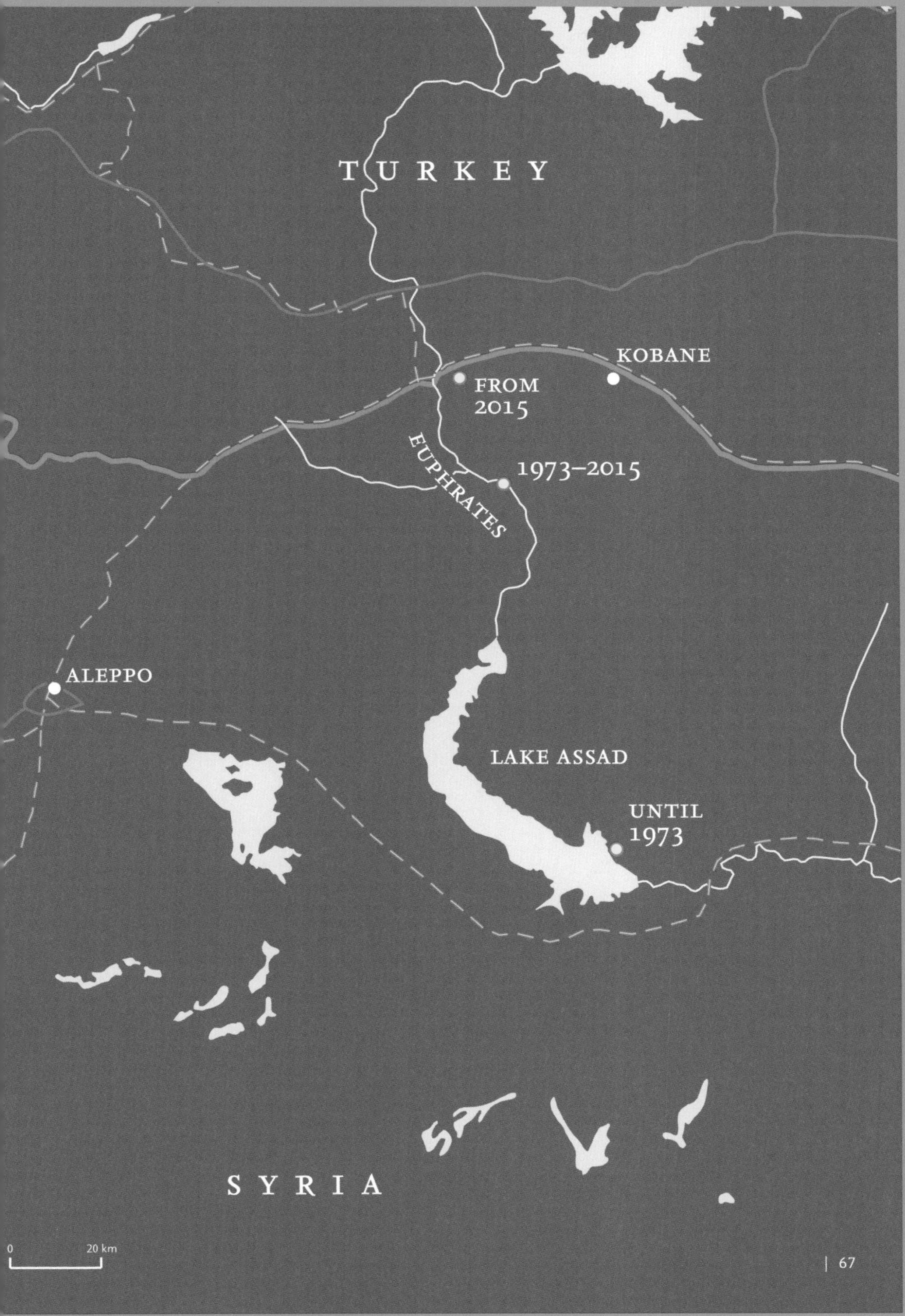

TURKEY

KOBANE

FROM 2015

EUPHRATES

1973–2015

ALEPPO

LAKE ASSAD

UNTIL 1973

SYRIA

0 20 km

Most of the world's enclaves are inhabited, though there are some that are used only for production, mainly agricultural. As well as these, there are some enclaves that have purely historical significance, with no other reason for their existence. This is the case with the **Tomb of Suleyman Shah**, a Turkish enclave in Syria.

Suleyman Shah was an important figure in Turkish history: he was the grandfather of Osman I, the founder of the Ottoman Empire, one of the greatest powers of all time. According to one legend, Suleyman Shah drowned in the Euphrates River though historians have not confirmed this. There is a possibility that a Suleyman Shah did drown in the Euphrates, though it was not the one who was grandfather to Osman I, but the founder of another Turkish state that was later integrated into the Ottoman Empire.

Nevertheless, it is generally accepted that the Tomb of Suleyman Shah was on a fortified hill, on the bank of the Euphrates, about a hundred kilometres south of the present-day Turkish–Syrian border. The border between Turkey and Syria, a former French colony, was established by a peace treaty between Turkey and France in 1921. This agreement stipulated that the land upon which the Tomb of Suleyman Shah stood (Qal'at Ja'bar hill, with an area of less than one hectare), should remain Turkish, with a small Turkish guard of honour.

However, as if the existence of such an enclave was not unusual enough, the creation of the large, artificial Lake Assad on the Euphrates River led to yet another strange situation. Due to the risk of the tomb being flooded, Turkey and Syria agreed to relocate the enclave seventy kilometres upstream, to just thirty-five kilometres from the Turkish–Syrian border. (Who knows why they did not decide to move the enclave to the border, so that it could be 'unified' with the rest of Turkey?) Anyway, until the beginning of 2015 and the escalation of the conflict in Syria, the tomb was located on a small peninsula beside the Euphrates. Then, because of the risk of the fighters of the Islamic State (ISIS) attacking that Turkish 'island' in Syria, the guard of honour was initially reinforced with additional Turkish special forces, accompanied by the threat from Turkey that it would defend its entire territory with its mighty army, including the Tomb of Suleyman Shah.

A few months later, Turkey changed its strategy: in a rapid military action, the body of Suleyman Shah was transferred again, the buildings on the peninsula were destroyed (so they would not fall into the hands of ISIS), and a new tomb was built in the north of Syria, between the Euphrates and the city of Kobane, only 200 metres from the Turkish–Syrian border. Turkish authorities have repeatedly stressed that the tomb has only been 'temporarily' relocated, and that it will be returned to its previous location once the situation in Syria stabilizes.

However, according to some statements by Syrian officials, that country does not recognize Turkey's right to independently decide on the location of the mausoleum and its exclave in Syria, and that moving the tomb in 2015 is a violation of the 1921 Treaty of Ankara on the Turkish-Syrian border.

SAINT PIERRE AND MIQUELON

CANADA | FRANCE

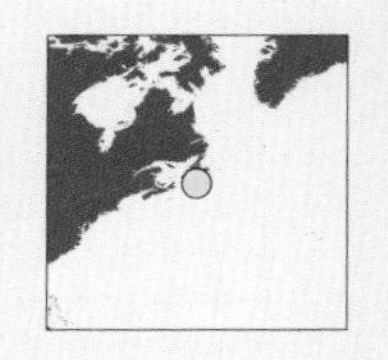

The enclave that is nearly 4,000 miles from its French homeland

46° 44' 00"N | 56° 08' 00"W

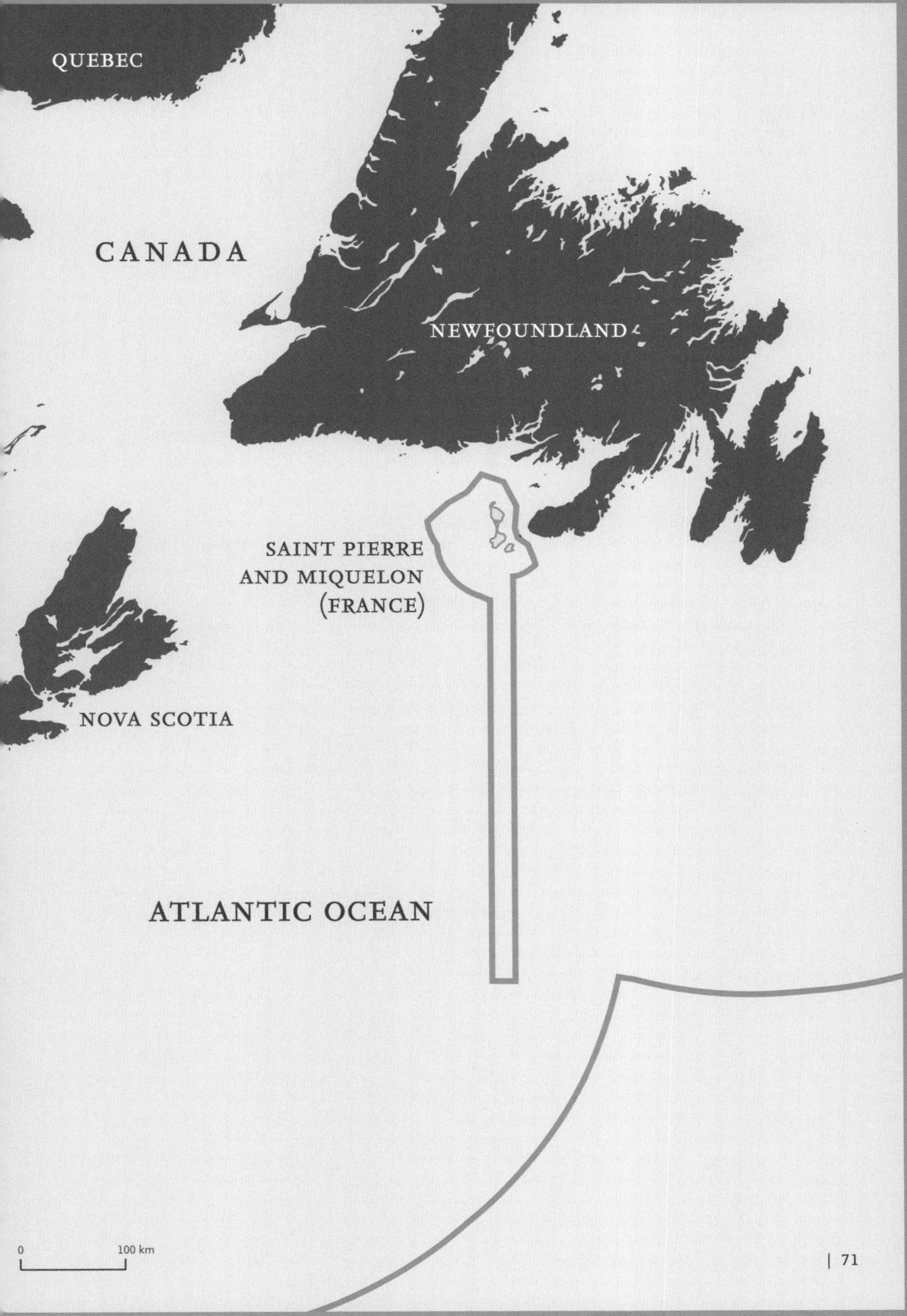

QUEBEC

CANADA

NEWFOUNDLAND

SAINT PIERRE
AND MIQUELON
(FRANCE)

NOVA SCOTIA

ATLANTIC OCEAN

0 100 km

Although most of the world enclaves are 'wedged in' somewhere between the land borders of their neighbouring countries, some enclaves are different. One such enclave is **Saint Pierre and Miquelon.**

Saint Pierre and Miquelon is a small archipelago, consisting of three main islands – Saint-Pierre, Miquelon and Langlade – and a large number of smaller islands. As Miquelon and Langlade are connected by a narrow sandbar, they are now considered to form a single island, so the name Miquelon is commonly used for both islands, shaped together as the number 8. The small archipelago is located twenty-five kilometres south of the Canadian island of Newfoundland, but politically and culturally it belongs to France, 3,800 kilometres away.

Saint Pierre and Miquelon, as the French overseas territorial collectivity, is an integral part of the French Republic and the European Union. The euro is used as a means of payment, though often, because of its location, both the Canadian and US dollars are also accepted. The archipelago, with a total area of about 240 square kilometres, has 6,000 inhabitants, who speak French (interestingly, this French is more similar to the language spoken in France than in the nearby Canadian Quebec). The main economic activities are fishing (although declining) and tourism, as the islands are easily accessible to visitors from Canada. In recent years, there has been some rising optimism that oil might be found in the surrounding waters, which could offer a significant boost to the archipelago's ailing economy.

What makes these islands a kind of enclave is the so-called 'exclusive economic zone' (EEZ) of Canada, which fully surrounds Saint Pierre and Miquelon. It was precisely on this issue that Canada and France had a long-running dispute in an international court, which ruled to assign the French archipelago a very unusually shaped EEZ (its boundaries resemble what many call 'the key'), so that French ships from international waters could reach the islands through a narrow corridor, nearly 200 kilometres long and only ten kilometres wide. However, Canada later exercised its right, under international regulations, to further extend its EEZ, so the 'French key' became a kind of maritime exclave of France in Canadian waters.

The possibility that significant oil reserves could be found in those waters is not likely to accelerate or facilitate the final defining of a compromise border between Canada and Saint Pierre and Miquelon, the last remnant of the once powerful colony of Nouvelle-France (New France).

Note: An exclusive economic zone is an area of coastal water and seabed within a specified distance of a country's coastline, to which the country claims exclusive rights for exploitation of the sea's biological and mineral resources (fish, oil and the like), including energy production from water and wind.

ISLAND ENCLAVES IN RIVERS AND LAKES

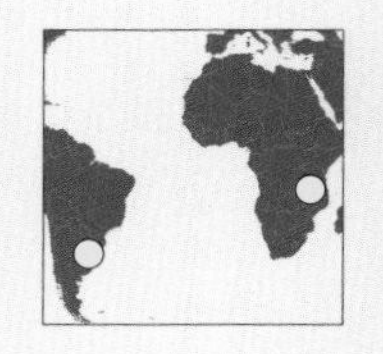

ARGENTINA | URUGUAY
ARGENTINA | PARAGUAY
MALAWI | MOZAMBIQUE

An archipelago in the Parana river: the largest true enclave in the world

Argentina | Uruguay 34º 10' 50"S | 58º 15' 05"W
Argentina | Paraguay 27º 31' 11"S | 56º 51 06"W
Malawi | Mozambique 12º 03' 02"S | 34º 40' 21"E

Martín García Island, Argentina | Uruguay

Many rivers and lakes form a natural border between countries. Occasionally, however, there is an unusual situation when a river or lake island belongs to one country even though it is located within the territorial waters of another, thus representing an enclave.

Martín García Island is an Argentinian island within Uruguayan territorial waters. It is located where the Uruguay and Paraná rivers meet to form the Río de la Plata estuary. According to the 1973 Agreement between Argentina and Paraguay, Argentina was to use this island as a nature reserve. Approximately 150 people live on less than two square kilometres. Interestingly, over time, the Uruguay River deposited huge quantities of sand and mud between the islands of Martín García and Timotéo Domínguez (part of Uruguay), causing them to merge into a single island with the only land border between the Argentine Republic and the Oriental Republic of Uruguay.

The **Filomena Islands** are a Uruguayan exclave within Argentinian territorial waters on the Uruguay River. The archipelago consists of several uninhabited islands, about 200 kilometres north of Buenos Aires, the capital of Argentina, and about 275 kilometres northwest of Uruguay's capital, Montevideo.

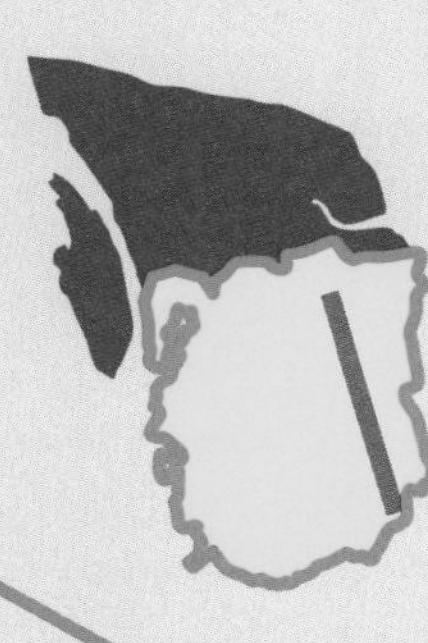
URUGUAY
MARTÍN GARCÍA
ISLAND
RÍO DE LA PLATA
0
1 km

The **Apipé Islands** form an archipelago of several Argentinian river islands completely surrounded by the Paraguayan stretch of the Paraná River. This archipelago consists of a few larger islands (Apipé Grande, 277 square kilometres; Apipé Chico, twenty-four square kilometres; San Martín, four square kilometres; Los Patos, twelve square kilometres) and many small islands. The two largest islands have a population of almost 3,000 people, while the other islands are uninhabited tourist sites. This archipelago is probably the world's largest true enclave.

Approximately forty-five kilometres upstream, there is another Argentinian island in Paraguayan waters: **Entre Ríos Island**. It is uninhabited, with an area of about thirty-five square kilometres. Nearby, there are several more islands in Paraguayan waters under Argentinian governance: **Caá Verá** (uninhabited, five square kilometres), **Verdes** and **Costa Larga**. The first two are located near the town of Itatí. Each of these islands varies in its size and shape, depending on whether mud and sand are being deposited or washed away.

Lake Nyasa (Lake Malawi) is probably the only lake with islands that are complete enclaves. **Likoma** and **Chizumulu Islands** are exclaves belonging to Malawi, but they are completely surrounded by territorial waters of Mozambique. The two islands form the Likoma District with an area of eighteen square kilometres and about 15,000 inhabitants. Owing to the mission established by the British on Likoma, these islands remained a part of Malawi, which was a British colony, instead of the nearby Portuguese Mozambique.

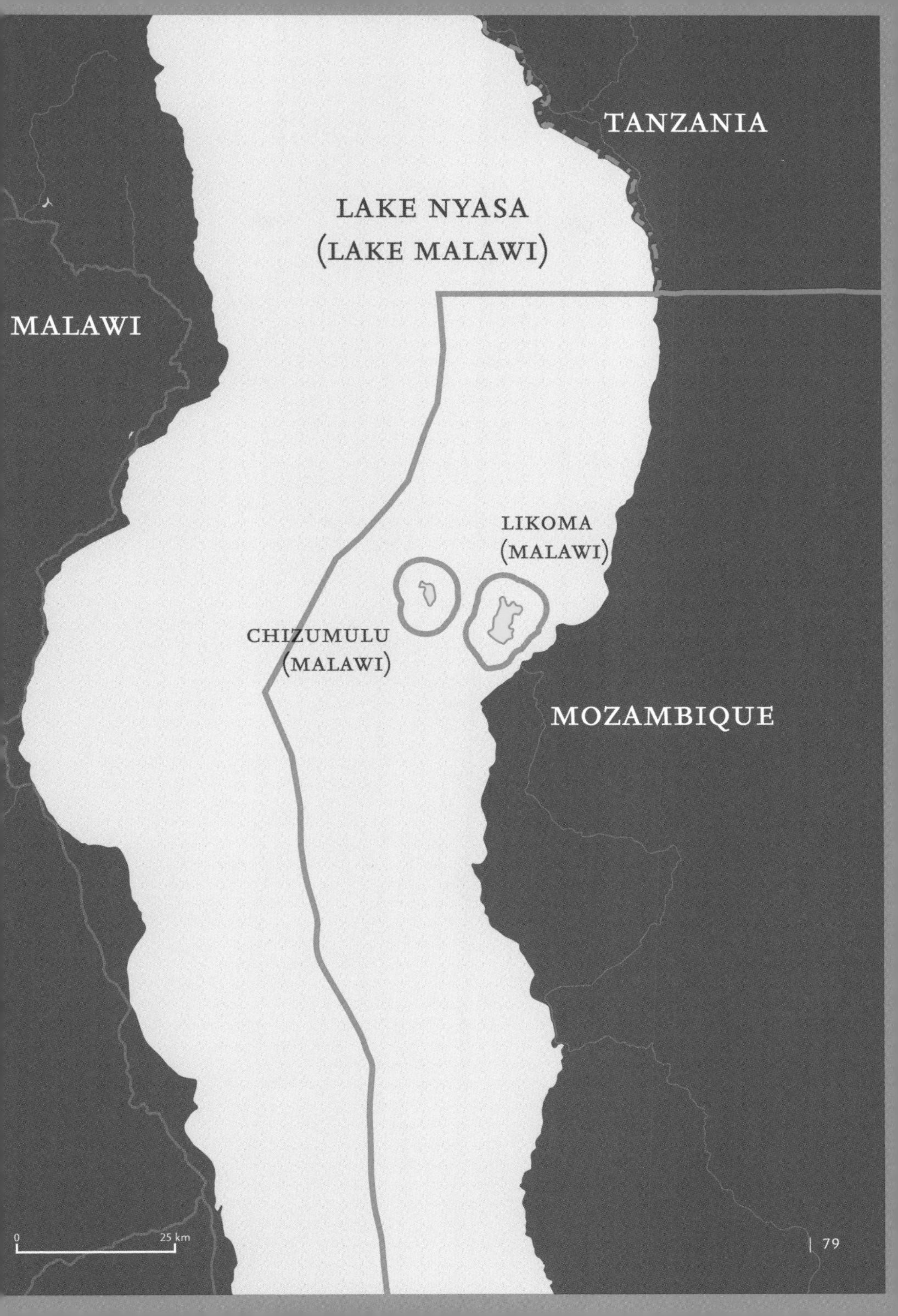
TANZANIA
LAKE NYASA
(LAKE MALAWI)
MALAWI
LIKOMA
(MALAWI)
CHIZUMULU
(MALAWI)
MOZAMBIQUE
0
25 km

CONDOMINIUM
PHEASANT ISLAND
MOSELLE RIVER

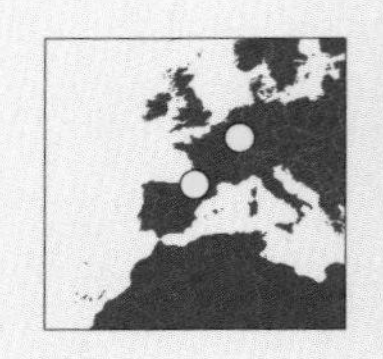

The island that has changed hands more than 700 times

Pheasant Island 43° 20' 33"N, 01° 45' 58"W
Moselle River 49° 28' 12"N, 06° 22' 04"E

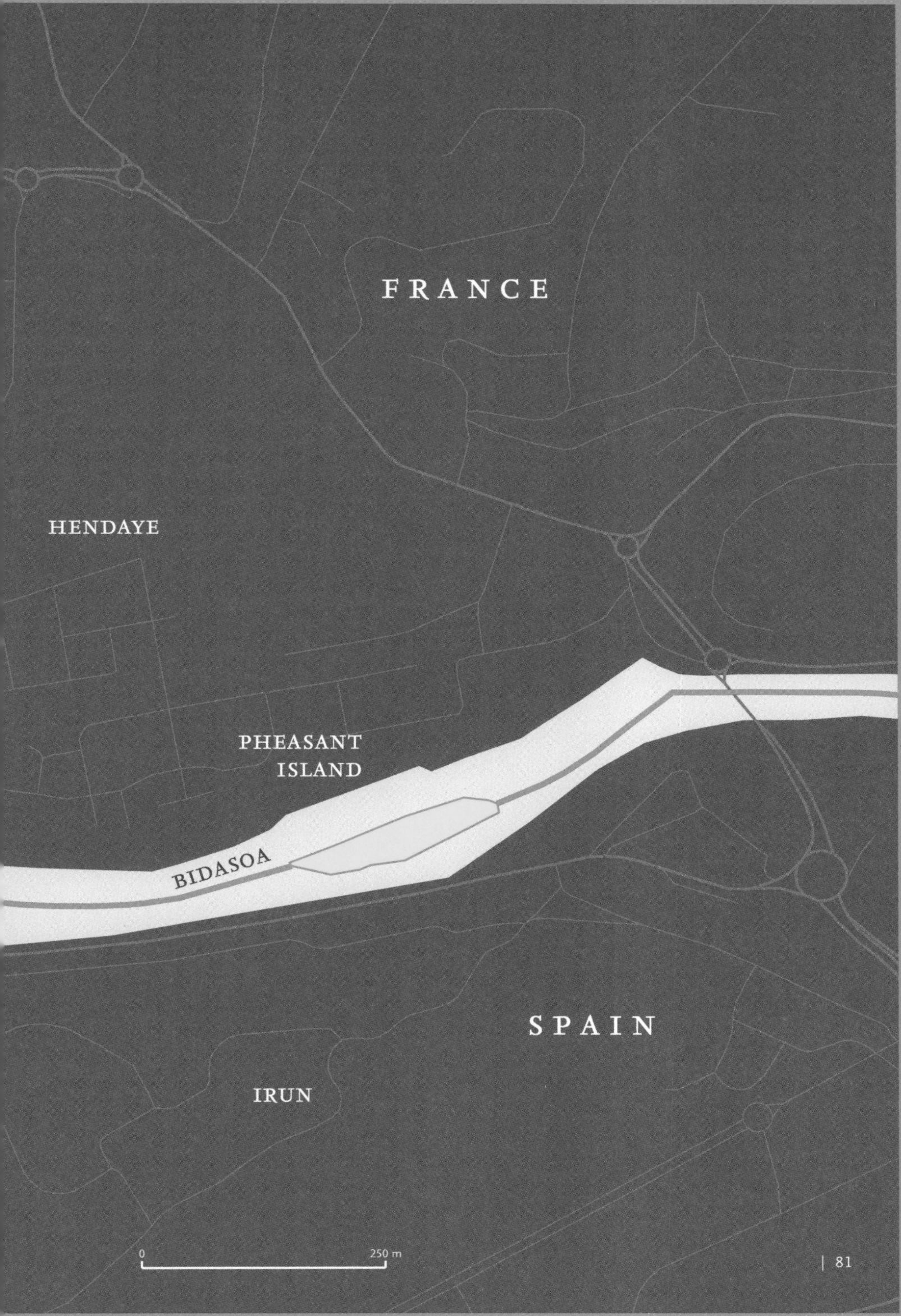

FRANCE

HENDAYE

PHEASANT
ISLAND

BIDASOA

SPAIN

IRUN

0 250 m

In international law, condominium (from the Latin *con-dominium*, 'joint ownership') means the joint management and authority of two or more states over a particular territory. Throughout history, there have been numerous examples of condominiums. With the formation of the modern nation-states, these territories mainly disappeared, by becoming part of one country or another. However, several condominiums do still exist.

Pheasant Island (Isla de los Faisanes in Spanish, Île des Faisans in French, Konpantzia in Basque) is a small uninhabited island – 6,800 square metres – on the River Bidasoa, about five kilometres upstream from where the river meets the Bay of Biscay (see map on p81). The island was established as a joint ownership (condominium) of France and Spain in their treaty of peace and demarcation signed in the second half of the seventeenth century. Under this treaty, which is still in force, the island became a very unusual condominium in that it is not, in fact, jointly owned by France and Spain, but ownership is shared. For half the year it belongs to the Spanish city of Irun, and for the remaining half, to its French twin-town of Hendaye. This means that over the past 350 or so years, the island has changed 'nationality' more than 700 times. To commemorate the signing of the treaty, a monument was built on the island, where it has the additional protection from the river waves. Throughout history, Pheasant Island was the location for many royal meetings and talks, but today, unfortunately, it is completely closed to visitors.

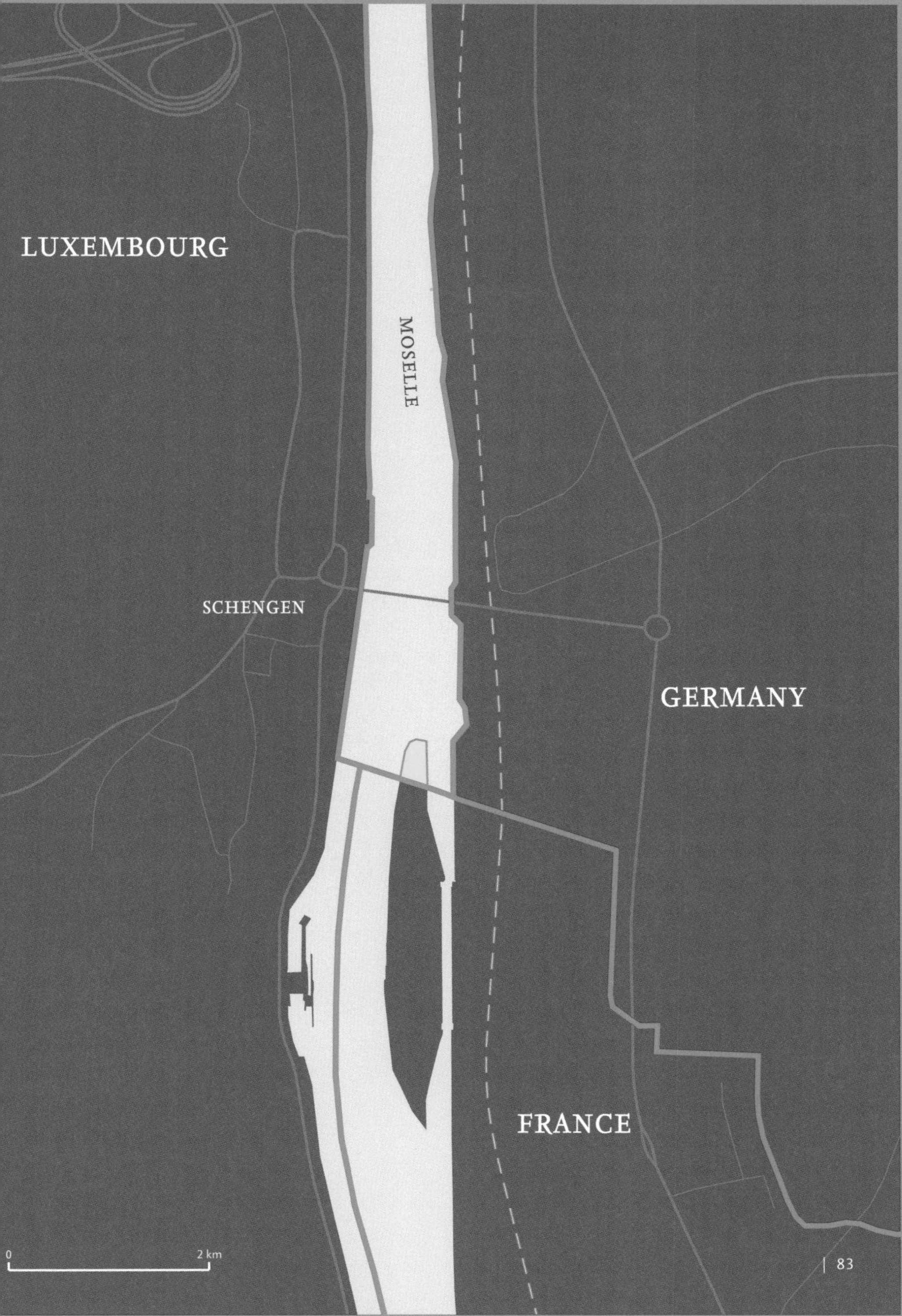

LUXEMBOURG

MOSELLE

SCHENGEN

GERMANY

FRANCE

0 2 km

According to the 1816 peace treaty between Germany and the Netherlands, the border between Germany and Luxembourg – a part of the Netherlands at that time – was established mostly along the **Moselle** (Mosel in German), Sauer (Sûre in French) and Our rivers. It was agreed that the rivers, bridges, dams and any islands (river islands are easily formed and demolished) would be under joint management. An island in the Moselle, located near the tri-junction of Luxembourg, France and Germany, mostly belongs to France, while its northern tip belongs jointly to Germany and Luxembourg (see map on p83). Between the two world wars, Germany suggested that the condominium and its division should be discontinued, but the small Grand Duchy boldly rejected the proposal from its powerful neighbour. The last amendments to the condominium agreement were related to the status of the bridges: Germany considered that the bridges should be divided in half, but the additional Germany–Luxembourg border agreement of 1984 stipulated that the bridges were also joint territory.

El Salvador, Honduras and Nicaragua have joint power over the **Gulf of Fonseca**, following a long conflict and negotiations, although the islands in the gulf are divided between El Salvador and Honduras.

Part of the **Paraná River** and the artificial lake created by the building of the Itaipu Dam is a condominium of Brazil and Paraguay.

The Republic of Bosnia and Herzegovina mainly consists of two political entities (Republika Srpska and the Federation of Bosnia and Herzegovina), but there is also **Brčko District**, which is formally a condominium of the two political entities although in practical terms it functions as a third entity.

If the negotiations between France and Mauritius are successful – they started off well, and then slowed down a bit – the world will perhaps get another condominium, **Tromelin Island** in the Indian Ocean. This low, flat island, with an area of only about one square kilometre, is located approximately 450 kilometres east of Madagascar and about 550 kilometres north of the French overseas territory of Réunion. Due to the demarcation lines, which were not clearly defined in the 1814 Treaty of Paris, both France and Mauritius claim sovereignty over the island. The reason lies in the abundance of fish in the waters belonging to Tromelin, as well as a potentially significant oil reserve. Nevertheless, France and Mauritius are close to reaching an agreement on joint management of the island.

Amazingly, there is an entire continent that could be regarded as a kind of condominium. The continent in question is **Antarctica**, jointly managed by about fifty countries, who were signatories to the Antarctic Treaty System.

Among the numerous condominiums that have existed in the past, many were quite distinctive.

One such example is **Cyprus**, which was a condominium of the Byzantine Empire and Umayyad Caliphate from the seventh to the tenth centuries. These two countries were almost continuously at war – except in Cyprus. During the whole time of joint rule, tax was duly collected and evenly divided into two equal shares.

The **Anglo-Egyptian Sudan** was another unusual condominium, having been created in an unusual way. Firstly, the United Kingdom occupied Egypt at the end of the nineteenth century, proclaiming it as a protectorate (although Egypt formally remained an autonomous region within the Ottoman Empire almost until the beginning of the First World War). Then the United Kingdom, together with its colony of Egypt, conquered Sudan (nowadays Sudan and South Sudan), and formally declared joint control, although the UK kept the predominant role throughout the duration of the condominium, which ended in 1955.

SPAIN–MOROCCO BORDER

The shortest land border in the world, separating two sovereign states

35° 22' 41"N | 03° 38' 15"W

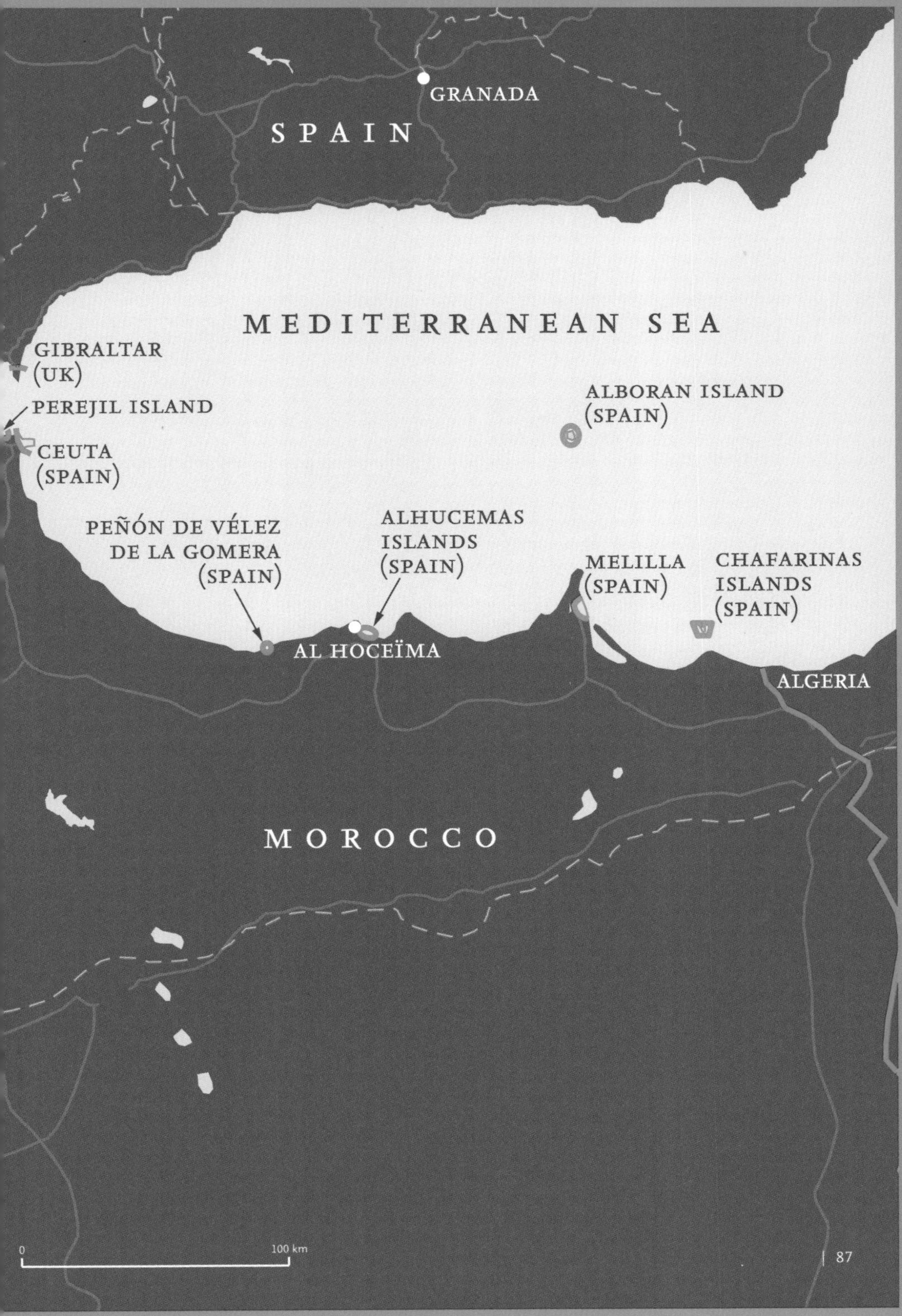

GRANADA
SPAIN
MEDITERRANEAN SEA
GIBRALTAR (UK)
PEREJIL ISLAND
CEUTA (SPAIN)
ALBORAN ISLAND (SPAIN)
PEÑÓN DE VÉLEZ DE LA GOMERA (SPAIN)
ALHUCEMAS ISLANDS (SPAIN)
MELILLA (SPAIN)
CHAFARINAS ISLANDS (SPAIN)
AL HOCEÏMA
ALGERIA
MOROCCO
0
100 km

British–Spanish relations have been cool for decades due to **Gibraltar**, a small rocky peninsula on the southern coast of the Spanish landmass. The United Kingdom has held this strategic peninsula, with its predominant control over the Strait of Gibraltar – the maritime connection between the Atlantic Ocean and the Mediterranean Sea – since as far back as the eighteenth century. Spain has often sought 'decolonization' of Gibraltar and its return to the Spanish crown, even though the Gibraltarians themselves oppose it.

This situation does not prevent Spain from holding on to a few of its own small 'Gibraltars' on the opposite side of the strait. Located on the north coast of Morocco, these are known as the **(Spanish) sovereign territories** (Plazas de soberanía). Their collective name refers to the fact that they have been a part of Spain ever since the formation of the modern Spanish state in the fifteenth and sixteenth centuries. The sovereign territories are fully Spanish and thus part of the European Union and the Schengen Area, but they have an undefined internal status in that they are not part of any Spanish province.

The sovereign territories are usually divided into major territories, which are Ceuta and Melilla, and minor territories, which refer to a few island and half-island enclaves. Currently, the Spanish term Plazas de soberanía is primarily used to refer to these minor territories, whereas Ceuta and Melilla are classed as autonomous cities.

In order to understand how Spain came into possession of these territories, it is necessary to dig into history. At the beginning of the eighth century, the powerful Islamic State, which extended over the whole of northern Africa and the Near East, conquered the largest part of the Iberian Peninsula, namely today's Spain and Portugal. The struggle for liberation from Muslim rule (the Reconquista) lasted a long time. The last Muslim state on the Iberian Peninsula, the Emirate of Granada, was reconquered at the end of the fifteenth century. After the withdrawal of the Muslim army to northern Africa, the rulers of Spain and Portugal followed, conquering several strategically important peninsulas and islands on the north shore of Morocco. They had two goals: to monitor the movements of the Muslim army and to prevent the frequent pirate attacks of North African Berbers on ships around the Strait of Gibraltar.

The sovereign territories today include a few towns, peninsulas and islands:

Perejil Island (Isla de Perejil, 'Parsley Island') is the most westerly of these territories. It is located within the Strait of Gibraltar, 250 metres off the coast of Morocco, eight kilometres from the Spanish town of Ceuta and almost fifteen kilometres from mainland Spain. Its original Berber name, Tura (meaning 'empty') is more applicable, as it is an almost completely bare, rocky island, measuring 480 metres by 480 metres. Even though this islet is generally of little significance, it was nearly a source of conflict between Spain and Morocco in 2002. Morocco tried to set up a small military base on Perejil Island ('in order to monitor illegal immigration'). This led to a rapid military response by Spanish Special Forces, the Navy and Air Force, who arrested the Moroccan soldiers and transported them to Ceuta, and then to the Moroccan border. It was agreed to return to the situation before Morocco's attempted occupation. Today the island is completely abandoned, and represents 'no man's land', closely monitored by both sides.

Only eight kilometres east of Perejil Island is the town of **Ceuta**. This Spanish enclave is located right across the strait from Gibraltar. It has an area of 18.5 square kilometres and almost 85,000 inhabitants. Ceuta has acknowledged Spanish governance since the second half of the seventeenth century, and today it has the status of an autonomous city. However, the Moroccan authorities consider all of these territories to be the remainders of Spanish colonial occupation and that they should be returned to the home country – Morocco – immediately. The vast majority of the Ceuta (and Melilla) population strongly opposes this idea.

Peñón de Alhucemas, Spain

Further east is **Peñón de Vélez de la Gomera** (peñón means rock). It is one of the Spanish rock forts found on the coast of northern Morocco. This peñón is about 120 kilometres southeast of Ceuta. It was a small, natural, rocky island until the 1930s, when a storm deposited large quantities of sand between the island and mainland, turning it into a specific type of peninsula, known as a tombolo or 'tied island'. The world's shortest land border is at that very point, only eighty-five metres long. The entire island measures 400 metres by 100 metres, making its area less than two hectares; it is uninhabited except for a small military base.

Further again to the east are the **Alhucemas Islands** (Islas Alhucemas). This small archipelago consists of three islands: Peñón de Alhucemas, Isla de Mar and Isla de Tierra. It lies 300 metres from the coast of the Moroccan town of Al Hoceïma, about 150 kilometres east of Ceuta and 85 kilometres west of Melilla. The total area of the archipelago is about 4.5 hectares. Peñón de Alhucemas is a small rocky island, approximately the size of two football pitches, and is occupied by a fort, a church and several houses. The archipelago has belonged to Spain since the mid-sixteenth century, when the local rulers gave it to Spain in exchange for help against the Ottoman Empire. Today, there is a small military garrison (twenty-five to thirty soldiers) on the island of Peñón de Alhucemas, while an even smaller military camp is located on the nearby Isla de Tierra, used for preventing illegal immigration.

Alboran Island (Isla de Alborán) is located in the Alboran Sea, the westernmost part of the Mediterranean, fifty kilometres north of the Moroccan coast and ninety kilometres south of the Spanish mainland. It has been a Spanish possession since 1540, with a small Spanish Navy garrison and an automated lighthouse on it. The island is a flat platform, slightly resembling a large aircraft carrier, with an area of over seven hectares. This is the only sovereign territory Morocco is not requesting to be returned.

The autonomous city of **Melilla** covers an area of about twelve square kilometres. Of its approximately 80,000 inhabitants, the majority are Spanish, with slightly fewer Berbers and some Jews. Morocco regards this area as an occupied territory as well, even though it has been under Spanish rule since the end of the fifteenth century. Today, Melilla is famous for its multiculturalism, in which Christians (about 55 per cent), Muslims (about 45 per cent) and small communities of Jews and Hindus live quite harmoniously.

The easternmost sovereign territories are the small **Chafarinas Islands** (Islas Chafarinas), consisting of three islands with a total area of about half a square kilometre. The islands are about three kilometres from the Moroccan coast, while the distance to the closest part of the Algerian coast is less than twenty-five kilometres. The most important of the three islands is the middle-sized one (covering fifteen hectares), Isabella II, on which there is a garrison with fewer than 200 soldiers. This archipelago has been under the Spanish crown since the mid-nineteenth century.

FRANCE–NETHERLANDS BORDER

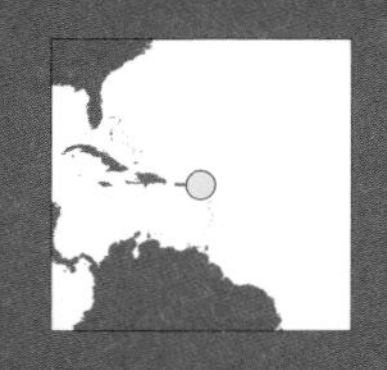

The European border that's an ocean away from the continent

18° 03' 51"N | 63° 04' 08"W

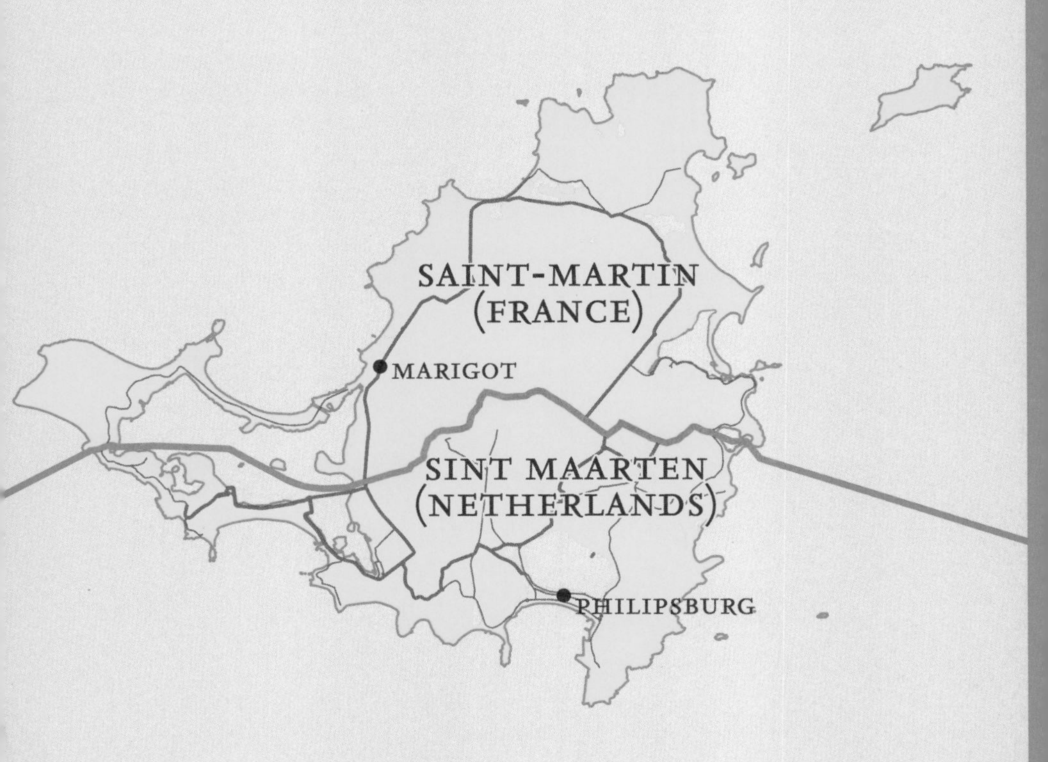
ANGUILLA
(UK)
SAINT-MARTIN
(FRANCE)
MARIGOT
SINT MAARTEN
(NETHERLANDS)
PHILIPSBURG
CARIBBEAN SEA
0
2 km

If you asked someone from Europe whether the Kingdom of the Netherlands and the Republic of France have a common border, it is likely that you would get the following answer: 'Of course not, Belgium is between them!' However, if you asked someone from the Caribbean the same question, it is possible that you would get the opposite response: 'Yes, of course they have a common border!'

How is it possible to get two contradictory answers? The explanation lies in the fact that both France and the Netherlands were significant colonial powers in the past, combined with the fact that the decolonization of their great empires did not apply to all colonies. This explains the border across the relatively small Caribbean island of **Saint Martin**.

The island is located in the northeastern part of the Caribbean, about 300 kilometres east of Puerto Rico, and has an area of about eighty-seven square kilometres. It has had a very turbulent history, and the Spanish, the English, the French and the Dutch have all ruled the island, or some of its parts, at some point. The island's current status, and its partition, is based on the agreement between France and the Netherlands that was signed in the middle of the seventeenth century. The northern part of the island (two-thirds of its area) belongs to the Republic of France and is known as **Saint-Martin**. The southern part (about a third of the island's area) is now part of the Kingdom of the Netherlands, going by the name of **Sint Maarten**. The population is roughly the same in both parts of the island, so the Dutch part, with around 41,000 inhabitants, is more densely populated than the French part, which has about 37,000 inhabitants.

Sint Maarten is one of the four constituent countries of the Kingdom of the Netherlands, and has the status of 'overseas country and territory' within the European Union (EU). Sint Maarten is a polyglot society: as per the 2001 census more than 65 per cent of the population were English speakers, almost 13 per cent were Spanish speakers and only around 4 per cent spoke Dutch. The guilder of the Netherlands Antilles is used as the method of payment, but it is soon to be replaced by the Caribbean guilder, although the US dollar is also widely accepted. The main basis of Saint Martin's economy is tourism, both by regular tourists and those who briefly visit the island during one of the numerous Caribbean cruises that sail around the islands. One of the biggest attractions in the Dutch part of the island is Maho Beach, where the runway of Princess Juliana airport is only separated from the beach by an ordinary fence and a narrow street. Due to the short length of the runway, aeroplanes have to fly very low over the beach. Watching the planes land from the beach is a popular and amusing pastime, which can also be very dangerous because of strong air turbulence caused by the planes as they pass overhead.

Saint-Martin is a French overseas community, occupying the northern part of this island. It is part of the EU, and the official currency is the euro. The main industry is tourism (85 per cent of the population is directly or indirectly involved in tourism), while most of the food and energy are imported, primarily from Mexico and the United States.

Nevertheless, Saint Martin is the richest territory in the Caribbean. Or at least it was until 6 September 2017, when Hurricane Irma hit the Caribbean islands – including Saint Martin – with 285 km/h (180 mph) winds causing extensive damage, injuries and deaths. The Dutch Red Cross estimated that nearly a third of the buildings in Sint Maarten had been destroyed and that over 90 per cent of the structures on the island had been damaged. The Netherlands and France sent additional police and military forces to stop widespread looting. As a consequence of the hurricane, Saint Martin's economy has suffered significantly.

Note: The Kingdom of the Netherlands consists of three island countries in the Caribbean (Sint Maarten, Aruba and Curaçao), plus the Netherlands. The Netherlands consists of twelve European provinces and three Caribbean special municipalities (Bonaire, Saba and Sint Eustatius). Officially, the status of all four countries is the same, although in reality this is often not the case, which is somewhat expected, since the European part of the Netherlands makes up about 98 per cent of the territory and population of the whole kingdom.

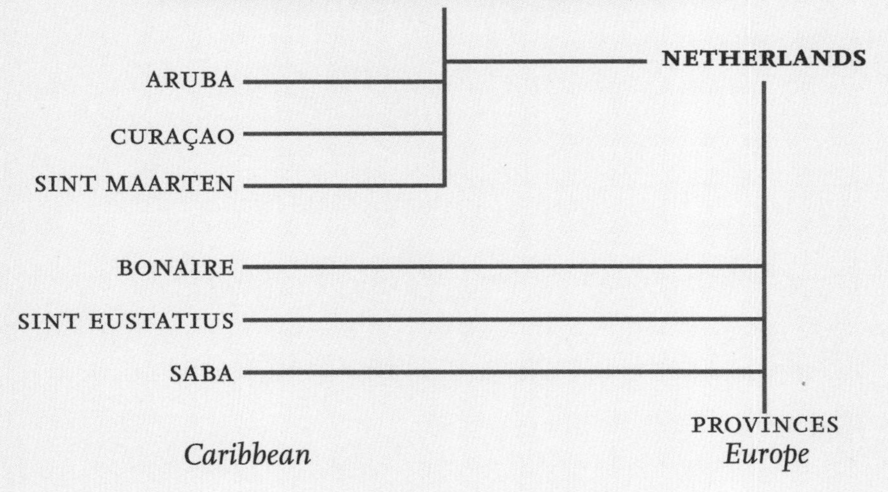
KINGDOM OF THE NETHERLANDS
NETHERLANDS
ARUBA
CURAÇAO
SINT MAARTEN
BONAIRE
SINT EUSTATIUS
SABA
PROVINCES
Caribbean
Europe

CANADA–USA BORDER

POINT ROBERTS
NORTHWEST ANGLE

Two countries brought closer together by a cultural venue that straddles the border

Point Roberts 48° 58' 09"N | 123° 03' 26"W
Northwest Angle 46° 49' 30"N | 56° 16' 30"W

Boundary Bay, Point Roberts, Canada | USA

The world's longest border, between the United States and Canada, at first glance seems quite straightforward: the Great Lakes and St Lawrence River make up a large part of it, and most of the rest follows a straight line. However, if we enlarge the map enough, we can see that this very long border (about 6,500 kilometres between mainland USA and Canada, plus another 2,500 kilometres between Alaska and Canada) has a few illogical and unusual points.

One of these points is the town of **Point Roberts**. It is located in the northwest of the United States, in the state of Washington. It is located on a peninsula that, with the exception of the southern part, belongs to Canada. Therefore, the land route from Point Roberts, on the south of the peninsula, to the rest of the United States leads entirely through Canada. This unusual situation is due to the demarcation along a line of latitude, by which all land south of the 49th parallel belongs to the United States, and all land north of that parallel to Canada. Of course, there are a few minor inconsistencies to this rule, which will be described below.

Point Roberts, with an area of about twelve square kilometres, has less than 1,500 inhabitants. The population grows in the summer to about 4,500, when many tourists from Canada go to their holiday homes there. In the town there is a primary school for children in lower grades. Older children with US citizenship take the school bus to Blaine, a city forty kilometres away, on the border between the United States and Canada, while Canadian children go to school in the nearby town of Delta, a suburb of Vancouver. Point Roberts has a small airport and port, enabling a direct connection to the rest of the United States, and telecommunications services are provided by companies from both the United States and Canada. Many residents of Point Roberts make a living from services provided to visitors from nearby Vancouver; a major attraction to tourists is the climate, which is milder and more pleasant than in the surrounding region.

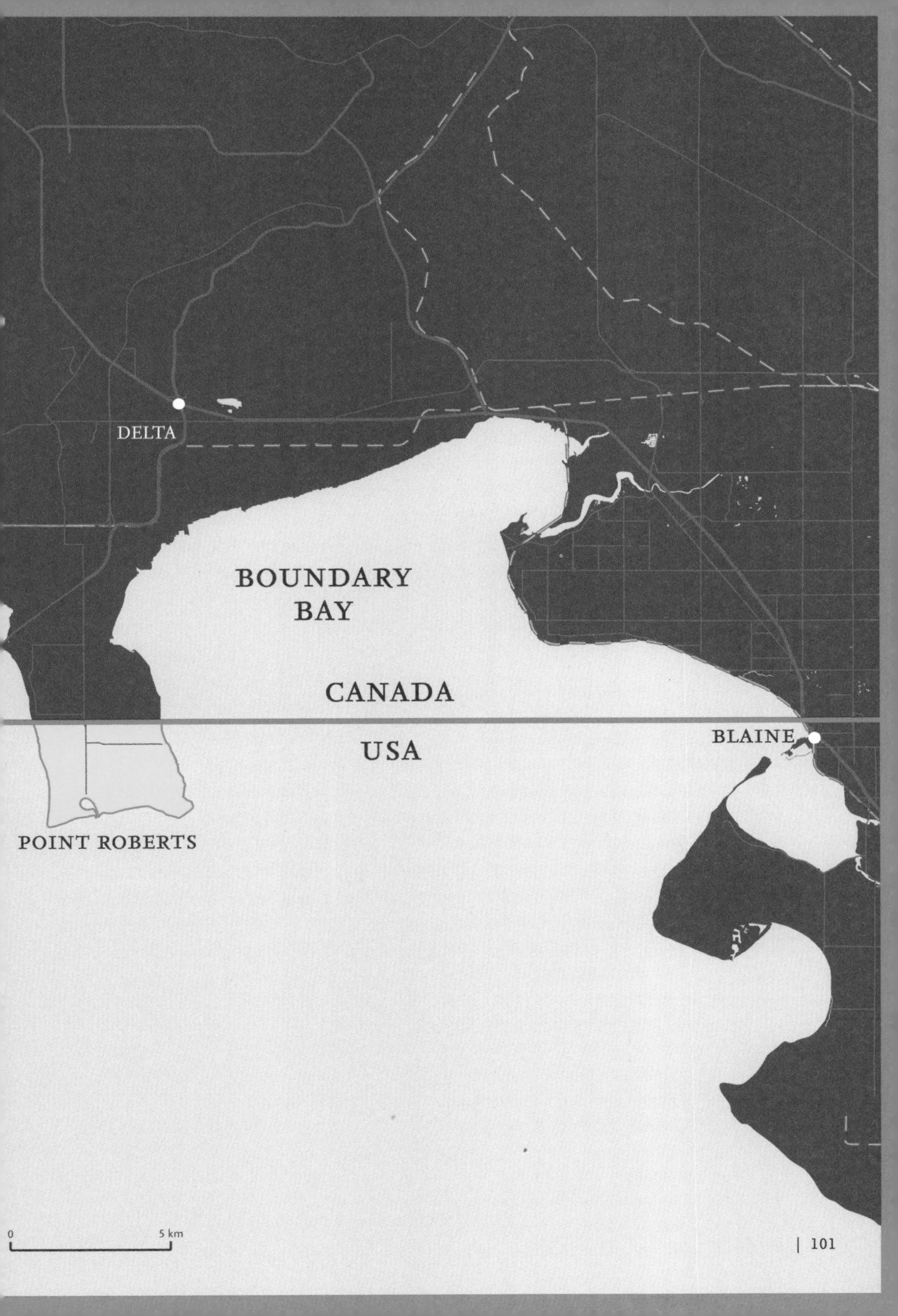
DELTA
BOUNDARY
BAY
CANADA
USA
BLAINE
POINT ROBERTS
0
5 km

Apart from Point Roberts, there are a few more unusual places on the border between the United States and Canada.

Northwest Angle, known simply as the Angle by the local population, is virtually the only territory of the United States, except Alaska, that lies to the north of the 49th parallel. As a result, the Angle, located on Lake of the Woods, is the northernmost location of the contiguous United States. The land area of this virtual exclave of the United States in Canada (the only land connection with the United States passes through Canada) is roughly 320 square kilometres, and its population is about 150. The border crossing between the Angle and Canada is also unusual: when someone crosses the border and enters the Angle from Canada, they are required to call US customs officials by videophone from a small booth; and vice versa, those visiting Canada from the Angle need to videophone their Canadian counterparts.

Eastwards from Angle along the Canada–USA border, there are numerous small peninsulas, which are de facto enclaves because they are separated from the homeland by lakes. One of these peninsulas is **Province Point** (one hectare, uninhabited), a peninsula in Lake Champlain, about 500 kilometres north of New York City.

About eighty kilometres east of Province Point there is a city divided by the border to form Canadian **Stanstead** and American **Beebe Plain**. Here we have two oddities. Firstly, the boundary line runs along the centre of the main street, so that the houses on one side of the street are in Canada and, on the other, in the United States. Secondly, there is a unique library and opera building (Haskell Free Library and Opera House), which opened at the beginning of the twentieth century. It was deliberately built on the border, in order to bring these two large countries culturally closer. Since almost all of the books and the stage are located on the Canadian side, it is often said that this is the only American library without books and the only American opera without a stage. The border is marked with a black line across the inside of the building.

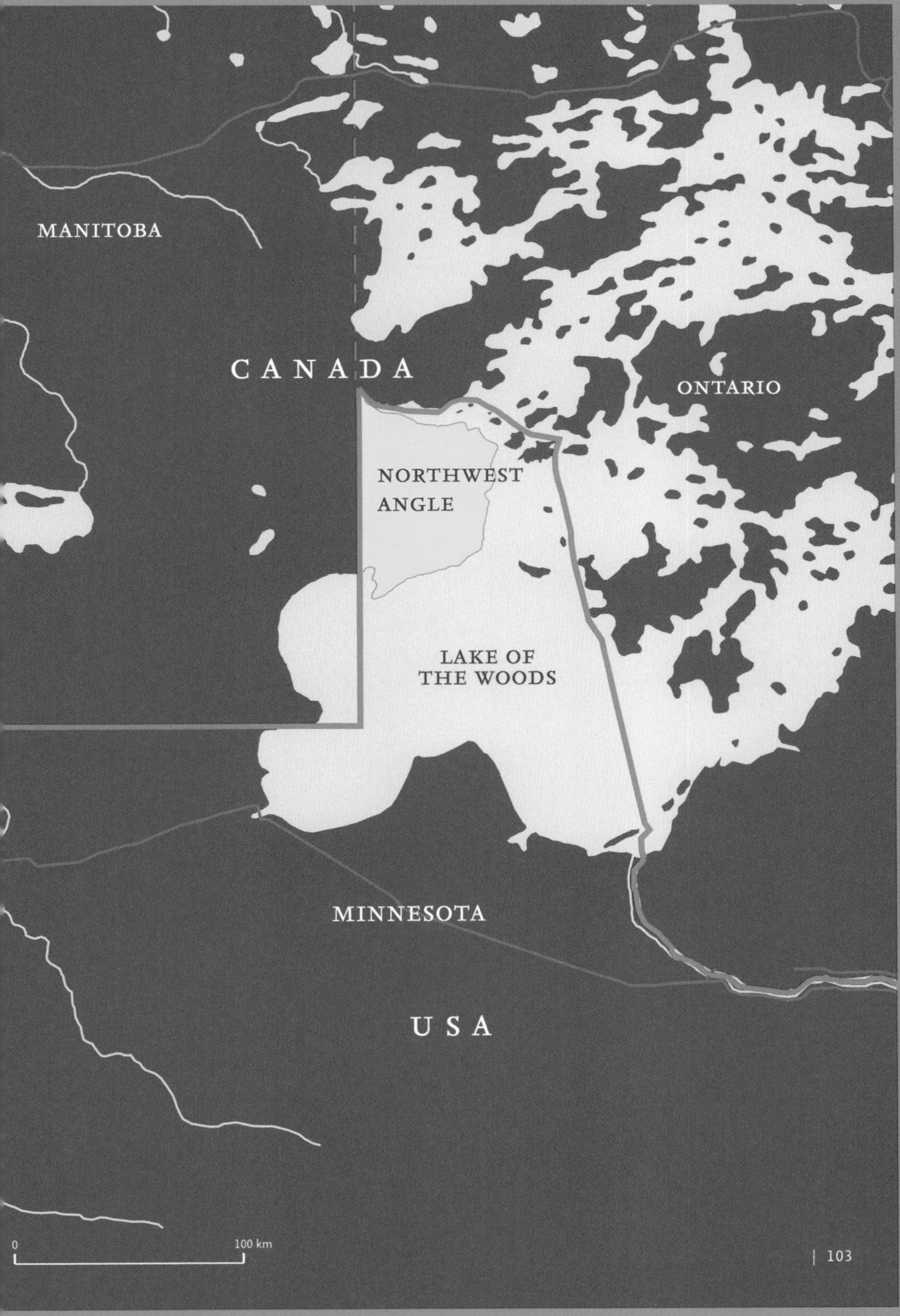

MANITOBA

CANADA

ONTARIO

NORTHWEST
ANGLE

LAKE OF
THE WOODS

MINNESOTA

USA

0

100 km

MÄRKET ISLAND

FINLAND | SWEDEN

The boundary that was redrawn to bring a new lighthouse into Finnish territory

60° 18' 01"N | 19° 07' 52"E

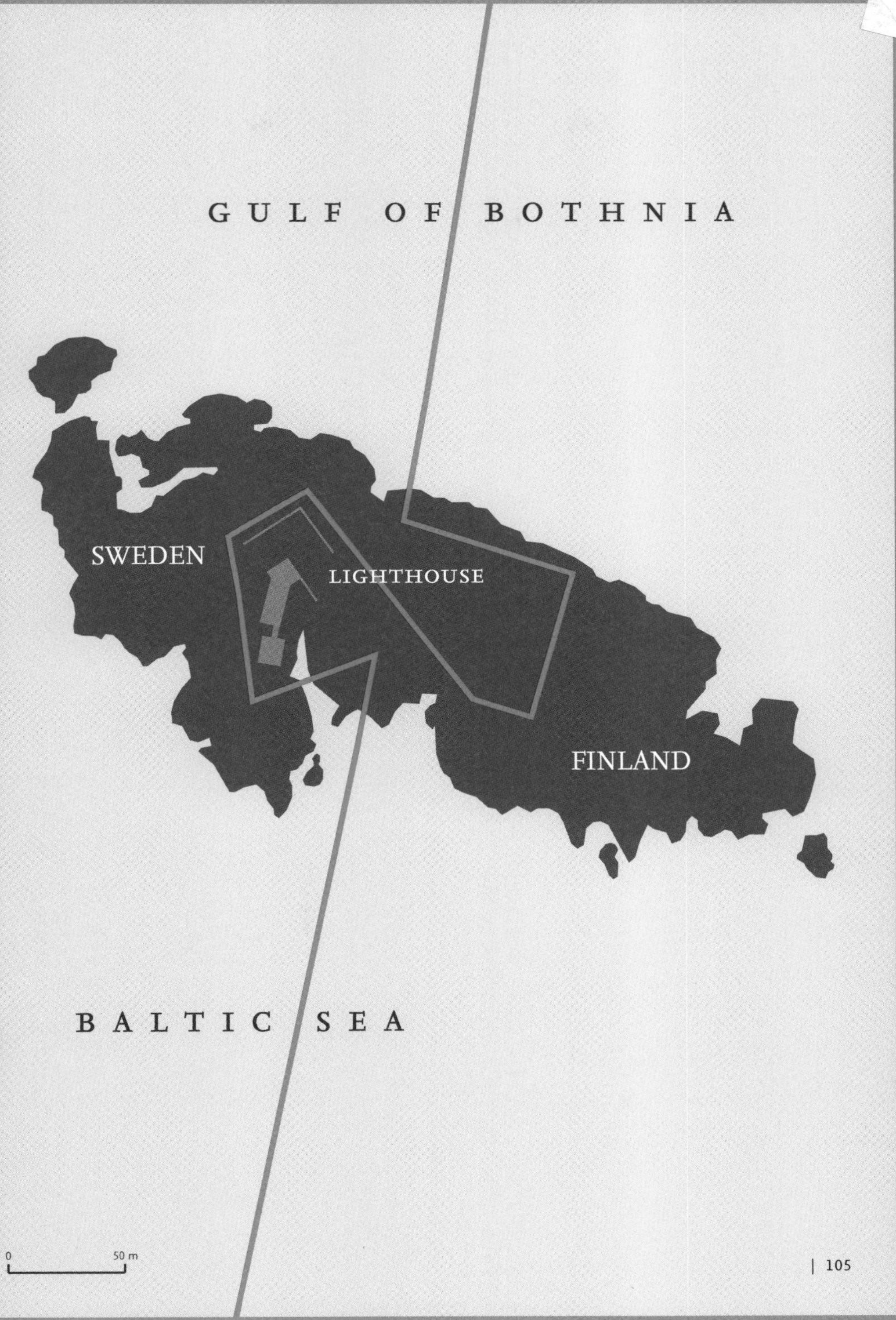

GULF OF BOTHNIA

SWEDEN

LIGHTHOUSE

FINLAND

BALTIC SEA

0 50 m

All those who are familiar with geography know that Finland and Sweden have a mutual land border far up in the north (where Santa lives and works, naturally). However, only a few will know that these two large Scandinavian countries border each other in the south too, just a few hundred kilometres to the north of the Swedish capital of Stockholm. There, at the entrance to the Gulf of Bothnia, lie the Åland (Ahvenanmaa) Islands, an autonomous province of Finland with a largely Swedish population. The westernmost island in this group is named **Märket Island** (meaning marker or marking, not a market place), and is distinctive because it is divided between Sweden and Finland and represents the only land border within the Åland Islands.

The fact that the island is divided between the two countries is not too strange – the border itself is much stranger, in terms of its shape. When one looks at the map, one can see that the island is divided by a line that looks like an inverted letter 'S'.

Why is there such an unusual division of this island? First of all, it should be noted that the island in question is rocky, inhospitable and unpopulated, covering an area of over three hectares or so, about 300 metres long and 150 metres wide. It is one of the smallest sea islands that is divided between the two countries.

At the end of the nineteenth century, the border ran straight across the centre of the island. At that time, Finland, as an autonomous Great Principality, still belonged to the Russian Empire, and the Russians and the Finnish decided to erect a lighthouse on this island. It wasn't until construction was fully completed that someone pointed out that the lighthouse had been built on the Swedish half of the island. That situation was finally resolved in 1985, by a complicated modification of what had been a straight and simple border across the island.

Two essential conditions had to be fulfilled when the border was redrawn:

- each of the countries had to end up with the same amount of land as before, and
- they also had to have the same length of coastline as before, so as not to interfere with the rights of the Swedish and Finnish fishermen in the waters encircling the island.

In this way, the land on which the lighthouse was situated was transferred to the governance of Finland. It is connected to the rest of the Finnish part of the island by just a narrow strip of land. It is interesting that Märket Island is also divided in terms of time, since Sweden and Finland are within different time zones.

DIOMEDE ISLANDS

RUSSIA | USA

The two islands are separated by less than four kilometres, and yet they are a day apart

65° 46' 54"N | 169° 03' 10"W

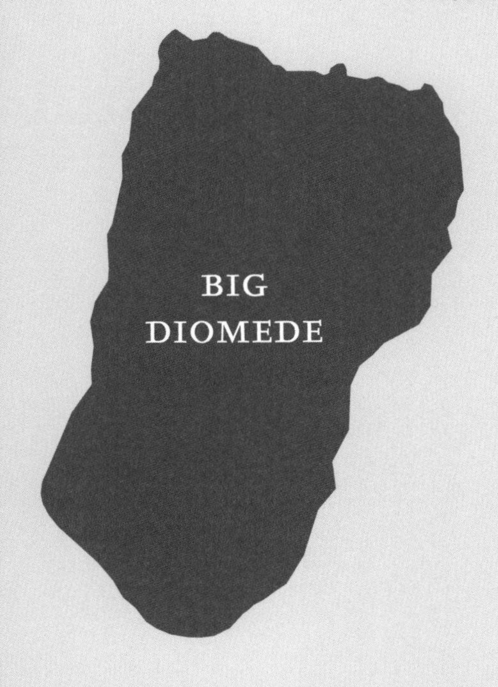
RUSSIA
USA
BIG DIOMEDE
DIOMEDE
LITTLE DIOMEDE
BERING STRAIT
0
2 km

While much of the Earth's surface is lined by various political boundaries, there is one boundary that divides the planet according to time. This is the International Date Line, an imaginary line that extends from the North Pole to the South Pole, separating one calendar day from the next. The line mostly passes through poorly populated areas of the Pacific Ocean, far from the inhabited islands and mainland. However, as it passes through the Bering Strait, where North America and Asia (that is, Alaska and Russia) are closest to each other, it creates an unusual situation for two islands – **Big Diomede** and **Little Diomede**.

Although, in geographical terms, they both belong to the same small archipelago, the Diomede Islands are politically separated by a boundary between the two world superpowers: Big Diomede belongs to Russia, whereas Little Diomede is a part of Alaska, and therefore the USA. Apart from this state border, there is another boundary that separates the two islands: the International Date Line. This very fact, combined with the fact that these islands are only about four kilometres apart, results in another oddity: on a clear day, from Little Diomede (also known as Yesterday Island), it is possible to see the next day on neighbouring Big Diomede (also known as Tomorrow Island). Regardless of the fact that these islands are only a short distance away from each other, the clocks on Big Diomede are twenty-one hours ahead of those on its smaller neighbour (although a difference of twenty-three hours might have been expected, this is not the case, due to the right of every state to define its own time zones). Roughly speaking, the difference between these two islands is a whole day.

Currently, nobody lives on the thirty-square-kilometre-island of Big Diomede, since the local population was forced to move to the mainland during the Cold War (to prevent contact with members of the same ethnic group on the neighbouring American island). Only a small military base is situated there today, meaning it is practically impossible to look back to the past. However, there are approximately 150 inhabitants on Little Diomede, which is just over seven square kilometres in area. These are predominantly the Chukchi people, the ethnic group inhabiting the Chukchi Peninsula in the far east of Russia. Life is certainly not easy on Yesterday Island, but it may improve in the future, if the bold idea of a bridge, tunnel or a combination of them is realized, connecting North America and Asia. These would definitely cross the Diomede Islands.

In any case, the very idea that, looking to one island from the other, it is possible to see another country, another continent and another day at the same time, sounds very unusual and amusing.

Diomede Islands, Russia | USA

MOROCCAN BERM

WESTERN SAHARA

The 2,700-kilometre sand wall that crosses the desert

23° 54' 00"N | 14° 32' 00"W

MOROCCO

CANARY ISLANDS
(SPAIN)

ALGERIA

TINDOUF

LAÂYOUNE

ATLANTIC
OCEAN

WESTERN
SAHARA
Administered
by Morocco

MOROCCAN BERM

MAURITANIA

0
200 km

There are various kinds of border in the world: land, river, maritime, open, closed... In certain situations, those who have claimed territory felt that it was not enough to draw an imaginary line around it and place border stones here and there, they needed to strengthen the border position with walls of varying sorts. Some of these walls are very well known. Probably the most famous one is the **Great Wall of China**, which marked the border between the Chinese civilization and the Mongolian barbarians. In addition to this, almost everyone has heard of the former **Berlin Wall**, and many people know about the walls (or fences) between the USA and Mexico, or between Israel and West Bank.

There is another 'wall' (a large earthwork barrier) that is little known, dividing a territory rich in phosphates, but which, in every other respect, has been exceptionally poor and constrained for decades. This is the **Moroccan Berm** (or Wall), which almost diagonally divides Western Sahara into two areas, one controlled by the Moroccans to the west and south, and the other to the east, which is controlled by the self-proclaimed Sahrawi Arab Democratic Republic. On the map, the berm is marked by a yellow line.

Western Sahara had been a Spanish colony until after the Second World War, when Morocco and Mauritania requested the retreat of the Spaniards to leave the region divided between them. Spain did not agree to retreat until the mid-1970s, which caused conflict with the Algerian-based Polisario Front, who claimed the territory. Shortly afterwards, Mauritania completely withdrew from the conflict, while the significantly superior Kingdom of Morocco relatively quickly took control of more than two-thirds of Western Sahara, including all major cities and the most important phosphate sites as well.

In order to secure the border, Morocco decided to construct a series of sand and stone structures with fortifications and ramparts, whose total length soon exceeded 2,700 kilometres. Alongside the berm, Morocco set up several million landmines and brought more than 120,000 heavily armed soldiers. In this way, the Moroccans ensured themselves unhindered access to the only natural resources in the west and south of Western Sahara – phosphates, and fish in the Atlantic Ocean. It is possible that there are potentially rich oilfields off the coast of Western Sahara, but the uncertain political status of the disputed territory may obstruct the exploitation of oil.

Today the Moroccan Berm represents a line between Moroccan-controlled areas and those areas controlled by the Polisario Front. Due to unfavourable living conditions, most of the Sahrawi – the local name for the indigenous population of this area – actually live in refugee camps in neighbouring Algeria. The berm extends from close to the southernmost point on Western Sahara's coast (the Polisario Front allegedly controls a couple of kilometres of the coast only, along the border with Mauritania), then it divides Western Sahara into two unequal parts, with a small section of the wall crossing the border into Mauritania, ultimately terminating in Morocco, close to the border with Western Sahara.

Although Morocco and the United Nations have signed a commitment to hold a referendum in Western Sahara on its future status, this has not happened so far and most probably never will. Taking into account, of course, that part of Western Sahara is extremely rich in phosphates and fish, and that it most probably contains rich oilfields, it is no wonder that Morocco refuses to surrender its power over this sparsely inhabited territory. In fact, about 90,000 Sahrawi people have remained in the area, and Morocco has already settled several hundred thousand of its own inhabitants there too.

PASSPORT ISLAND

BAHRAIN | SAUDI ARABIA

The island that was created to connect two countries

26º 11' 01"N | 50º 19' 28"E

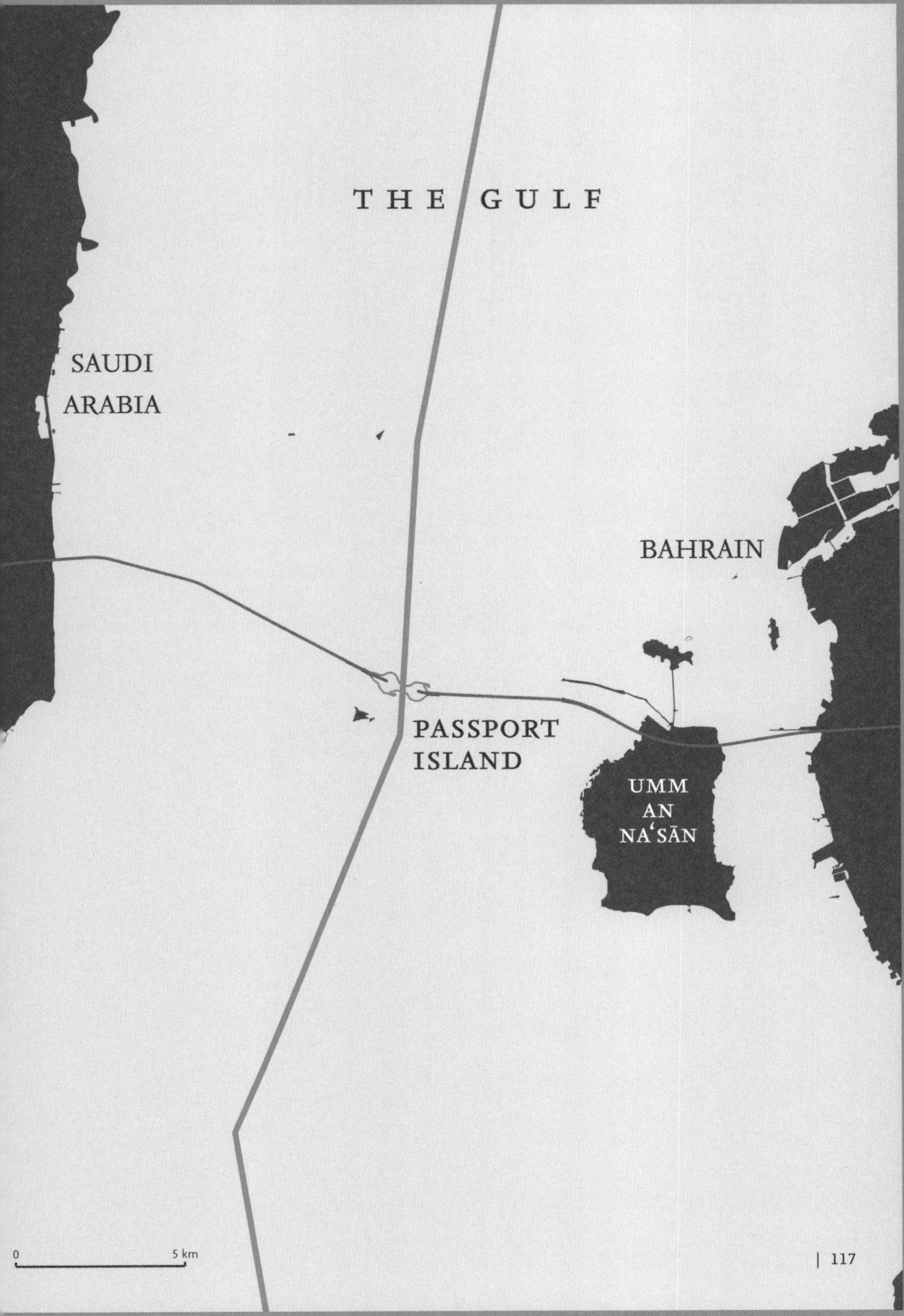

THE GULF

SAUDI ARABIA

BAHRAIN

PASSPORT ISLAND

UMM AN NAʿSĀN

0 5 km

Is it possible for an island nation, which does not share any islands with any of its neighbours, to still have a land border? The answer is yes, at least when it comes to Arabian countries extremely rich in oil and gas.

The idea of linking Saudi Arabia and its little neighbouring island of Bahrain was born long ago. It was publicly presented for the first time in the mid-1950s, when the King of Saudi Arabia visited his counterpart, the Bahrain Hakim (sheikh). A more concrete outline of this idea started to appear in the late 1960s, and construction of the bridge (now known as the King Fahd Causeway) took from 1981 to 1986.

The bridge represents a significant construction achievement. It consists of a system of bridges, with a total combined length of approximately twenty-five kilometres, while the four-lane road is over twenty metres wide. Large quantities of stone, concrete and reinforced steel were used in its construction. The entire causeway consists of three segments: firstly, the segment from Bahrain to the island of Umm an Na‘sān; secondly, from Umm an Na‘sān to the border station on **Passport Island**; and thirdly, from the border station to the Saudi mainland.

Passport Island? No, this is not a natural island, but a very large artificial one, almost two-and-a-half kilometres long and more than half a kilometre wide. The island was made in the shape of the figure '8', where one extension belongs to Saudi Arabia, and the other belongs to Bahrain. The island houses border stations for both Saudi Arabia and Bahrain, along with the buildings of the Causeway Authority, two mosques, the coast guard towers of both kingdoms and two tower restaurants (approximately sixty-five metres high), as well as one inevitable McDonald's on the Saudi side of the island. Otherwise, the island is enriched with greenery, above all with beautiful grasslands and lines of palm trees, so the impression is that of a natural island.

There are plans for the further expansion of this island in order to increase the number of traffic lanes. The plan includes construction of a large commercial centre on the Bahrain side, with a number of restaurants, coffee shops and stores, and even a well-equipped medical clinic.

NORTH SENTINEL ISLAND

INDIA

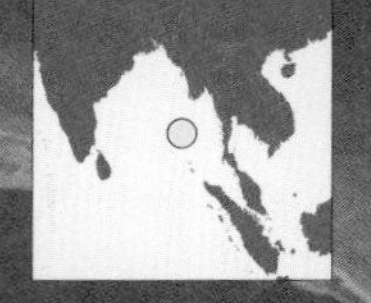

The island whose inhabitants choose to be cut off from the rest of the world

North Sentinel Island, India

11° 34' 20"N | 92° 14' 38"E

MYANMAR
(BURMA)
INDIA
BAY OF
BENGAL
NORTH
ANDAMAN
ANDAMAN
ISLANDS
MIDDLE
ANDAMAN
SOUTH
ANDAMAN
PORT
BLAIR
NORTH
SENTINEL
ISLAND
ANDAMAN
SEA
LITTLE ANDAMAN
0
50 km

Some borders are clearly visible, some are reinforced with walls or fences, but there are others that are invisible, because their neighbouring countries wish to make life easier for their citizens. There are also borders that no one officially acknowledges, even though they very much exist. One such example surrounds the small island of **North Sentinel**.

North Sentinel belongs to the Andaman Islands, an Indian archipelago in the Bay of Bengal. The archipelago is located opposite the shores of Myanmar (Burma) and, together with the southern archipelago of the Nicobar Islands, forms the Andaman and Nicobar Islands (A&N), one of the seven union territories of India. The diversity of these islands is huge, ecologically, ethnically, linguistically and economically.

However, North Sentinel Island, although only fifty kilometres away from Port Blair, the capital of A&N, stands out from all other islands of this union territory – this is the only island that is home to a tribe that absolutely refuses to have any contact with the rest of the world. The island has an area of about seventy square kilometres; it is almost completely forested, relatively flat, and has a population of between 100 and 500 Sentinelese people. It is not known by what name these people call themselves; according to unconfirmed sources, they call their island Chiö-tá-kwö-kwé. In all likelihood, their language is completely incomprehensible to the inhabitants of the nearby islands.

It is assumed that the Sentinelese are the direct descendants of the first group of people who left Africa about 60,000 years ago and headed for Australia, following the southern coast of Asia. They are short, dark-skinned and have curly hair. As regards their general level of technological development, they are still at the level of the Stone Age: they probably do not have any knowledge of fire, agriculture, or numbers greater than three, and survive only from hunting and gathering. Their main sources of food will be wild boars, fruit, fish, crabs, honey, and turtle and seagull eggs. They use bows and arrows, spears and harpoons for hunting. They have some metal weapons and tools, which they will have obtained from shipwrecks around the island.

The Sentinelese are extremely hostile to outsiders – almost all previous attempts to contact them have ended tragically. This tribe does not want any contact with the outside world, and there are unconfirmed reports that they killed several fishermen who had been accidentally stranded on the coral reefs around the island, plus at least one Christian missionary who recently tried to convey the word of God to them. All Indian expeditions have had a similar experience: their boats and helicopters were shot at.

All this has led to the establishment of a kind of border around the island of North Sentinel. Although the island is formally a part of the union territory of A&N, there is no form of government on the territory itself nor from India. Thus, North Sentinel can be regarded as an autonomous region of India or even as an undefined state under the protection of India. The A&N authorities officially stated in 2005 that they no longer intend to establish any contact with the Sentinelese and that they would let them live completely undisturbed by the outside world. Furthermore, a decision was made to discourage any approaches to the island, partly so that the Sentinelese would not attack any visitors, but also to prevent the visitors from transmitting any potentially dangerous diseases to them. A restricted zone was established, which extends five kilometres around the island.

The 2004 earthquake and tsunami had unexpected consequences for North Sentinel: the island rose by almost two metres, becoming attached to the neighbouring islet. The raised coral reef created large, partially closed lagoons as well as new dry areas. The island was thus significantly enlarged, but there is a possible danger that the previous lagoons, the main fishing areas of the Sentinelese, have become dry. According to available data, most of the Sentinelese survived, and a few days after the earthquake, an Indian helicopter was driven away by fired arrows and spears.

On the other side of the world, on the border of Brazil and Peru, there is another unofficial indigenous state: **Javari Valley** (Vale do Javari). This is a territory slightly smaller than Portugal, inhabited by numerous tribes with about 3,000 people. At least fourteen of these tribes have virtually no contact with modern civilization.

BORDER QUADRIPOINTS

NAMIBIA | ZAMBIA | ZIMBABWE | BOTSWANA
LITHUANIA | POLAND | RUSSIA
CANADA
CATANIA PROVINCE, SICILY

A quadripoint is a geographical point at which the borders of four regions intersect

Namibia | Zambia | Zimbabwe | Botswana 17° 47' 13"S, 25° 15' 41"E
Lithuania | Poland | Russia 54° 21' 49"N, 22° 47' 30"E
Canada 59° 59' 43"N, 102° 00' 59"W
Catania province 37° 44' 46"N, 15° 00' 57"E

Four Corners Monument, USA

A **tripoint** or **tri-border area** is a geographical point at which the borders of three regions intersect. These regions may be countries or their constituent elements. There are currently between 150 and 200 country tripoints in the world. Some of them are not precisely defined, because the borders of the countries around them are not fully defined, while some are situated in waters (seas, rivers or lakes). Land tripoints are usually clearly marked, sometimes even with a suitable pillar, or, as in the case of the tripoint of Austria, Slovakia and Hungary, a three-sided table, where each side of the triangle bears the coat of arms of one of the countries.

We may occasionally hear that an event occurred at the tripoint of countries X, Y and Z. But what about **quadripoints**? Have we ever heard that something happened at the quadripoint of countries A, B, C and D? Probably not, because the waves of the African Zambezi River are not significant enough to make the news.

Why the waves of the Zambezi? Currently, on the whole of the planet, with its nearly 200 sovereign countries and about a hundred other territories of various levels of sovereignty, one would expect there to be a fair number of quadripoints – but that is not the case. There may be one quadripoint, which happens to be in the middle of the Zambezi River, where the borders of **Namibia, Zambia, Zimbabwe and Botswana** meet at one point. The problem with this quadripoint is that Botswana and Zambia do not recognize it, claiming that two tripoints (Namibia–Zambia–Botswana and Zambia–Zimbabwe–Botswana) only 100–200 metres apart, are what exist instead. These two countries already have a plan to build a bridge over their short border between the two tripoints. So far, Namibia and Zimbabwe have not objected, which means they probably agree, at least tacitly, that there is no quadripoint after all.

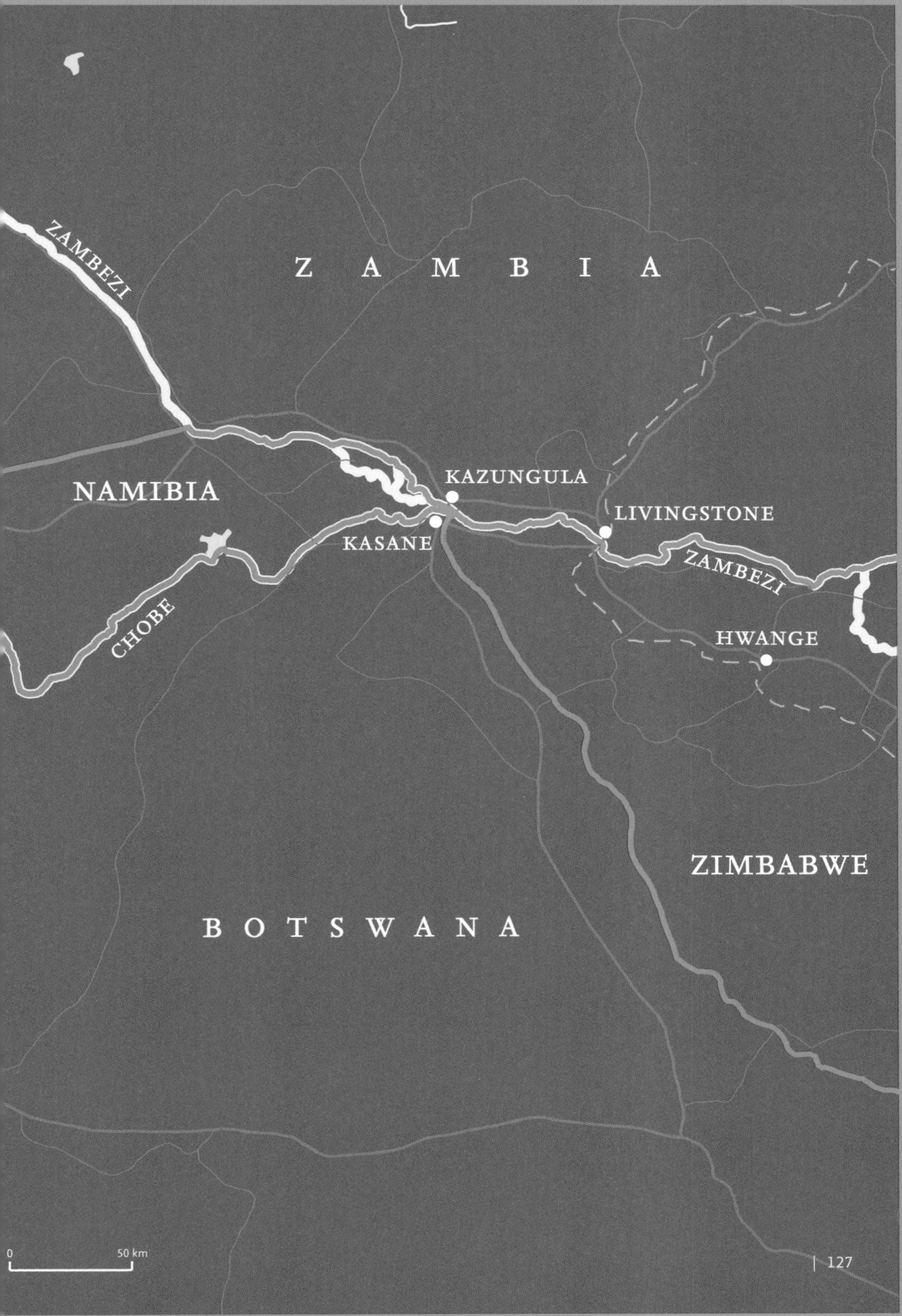

ZAMBEZI
ZAMBIA
NAMIBIA
KAZUNGULA
LIVINGSTONE
KASANE
CHOBE
ZAMBEZI
HWANGE
ZIMBABWE
BOTSWANA
0 50 km

There are other kinds of quadripoints in the world, though. One example is the quadripoint of **Austria and Germany**, near the Austrian pene-enclave of Jungholz; (see map on p36) and there are similar quadripoints between the **Netherlands and Belgium**, in the divided town of Baarle. (see map on p32) Such quadripoints are known as **binational quadripoints**.

In addition, there are **combined quadripoints** – places where the borders of, for example, two countries and two provinces of a third country intersect at one point (or any such similar combination). Such is the example of the quadripoint of **Lithuania, Russia (Kaliningrad) and two Polish provinces**.

In recent history, there were two quadripoints. One of them existed for several months in 1960, in the southern part of Lake Chad, where three independent countries (Nigeria, Chad and Cameroon) met at one point with British Cameroon. After a referendum on this British mandate, the northern part decided to merge with Nigeria, and the southern part with Cameroon, turning the quadripoint into a tripoint. The second, similar quadripoint lasted much longer, from 1839 to 1920. It was formed by three independent countries (Belgium, the Netherlands and Prussia, namely Germany) and Neutral Moresnet. Moresnet was a condominium, established by the Netherlands and Prussia. After Belgium separated from the Netherlands, the condominium was governed by Belgium and Prussia, while the unification of Germany in 1876 led to Moresnet being governed by Belgium and Germany (although Moresnet slowly obtained wider internal governance, and complete independence was increasingly demanded). Under the Treaty of Versailles in 1919, Neutral Moresnet, along with several other nearby German towns, was awarded to Belgium. Today, these territories represent part of the German-speaking Community of Belgium, the smallest federal community in Belgium.

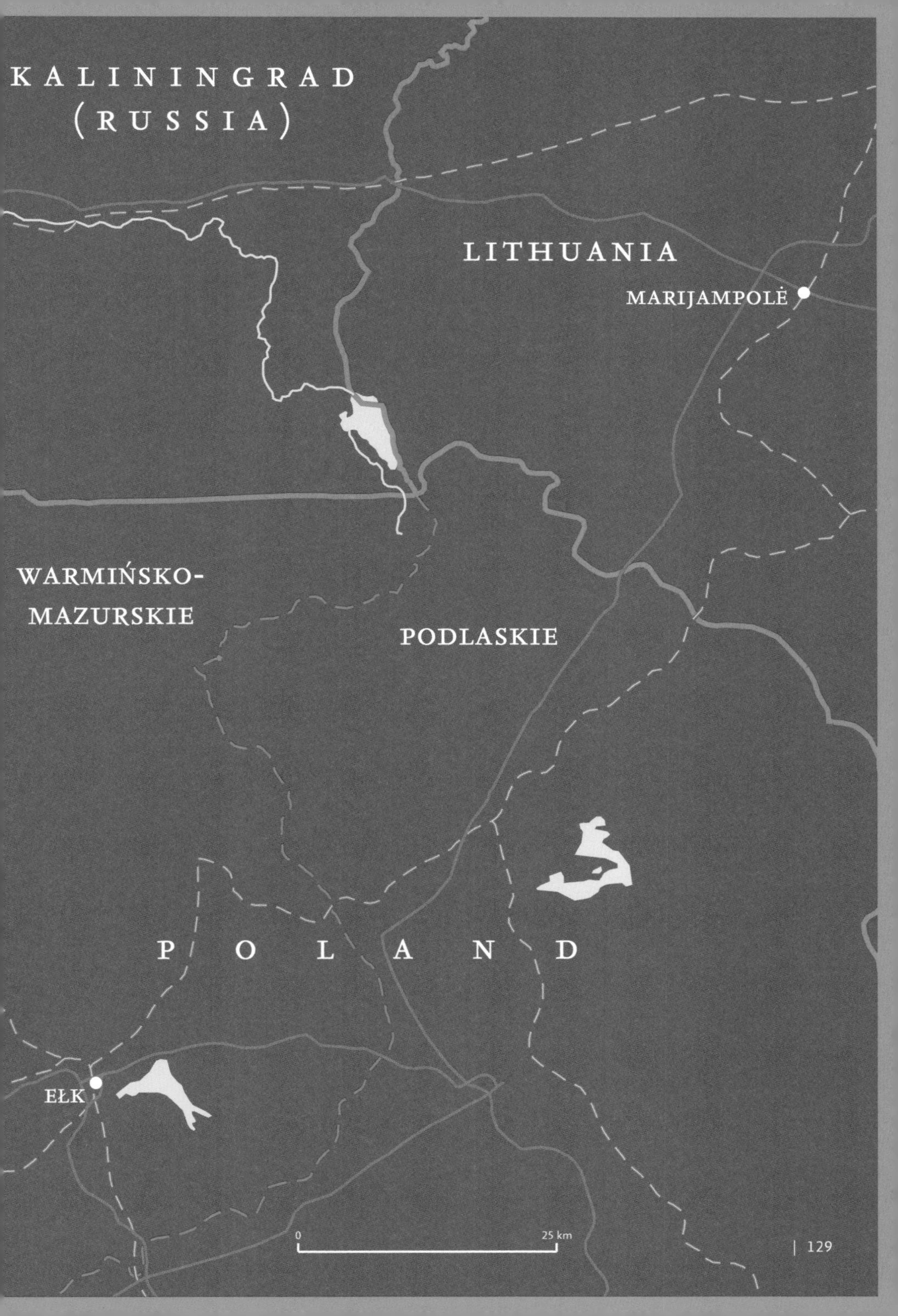
KALININGRAD
(RUSSIA)
LITHUANIA
MARIJAMPOLĖ
WARMIŃSKO-
MAZURSKIE
PODLASKIE
POLAND
EŁK
0
25 km

Quadripoints exist among the constitutive parts of various countries, the most famous being the quadripoint of the states of **Colorado, Utah, New Mexico and Arizona** in the USA. This is marked by an appropriate monument, known as the Four Corners Monument. The existence of another North American quadripoint between **Manitoba, Saskatchewan, Northwest Territories and Nunavut** in Canada has not yet been officially confirmed. Across the Atlantic Ocean, in the UK, there is a monument to the former quadripoint of the four English counties of **Gloucestershire, Oxfordshire, Warwickshire and Worcestershire**, known as the Four shire stone. Since the changes to the Worcestershire boundary in 1931, this monument has stood on the tripoint of the other three counties.

GREENLAND
(DENMARK)

NORTHWEST
TERRITORIES

NUNAVUT

ANADA

MANITOBA

SASKATCHEWAN

USA

0 500 km

At a lower administrative level, for example at the level of cities and municipalities, there are also **multipoints**, some of which have considerably greater complexity. In Finland, near the city of Turku, there is a **seven-point**, where seven municipalities meet (for a while it was an eight-point, until two of the municipalities merged). In Florida (USA) there is a local **five-point**, while in the Philippines there is a **six-point** and an **eight-point**. Ireland and Italy both have **ten-points**. Out of the twenty municipalities that make up Park Etna in Catania province in Sicily, Italy, as many as eleven (including one of them twice) border on one point, which is (probably) the only **eleven-point** in the world, located as it is in the centre of Mount Etna.

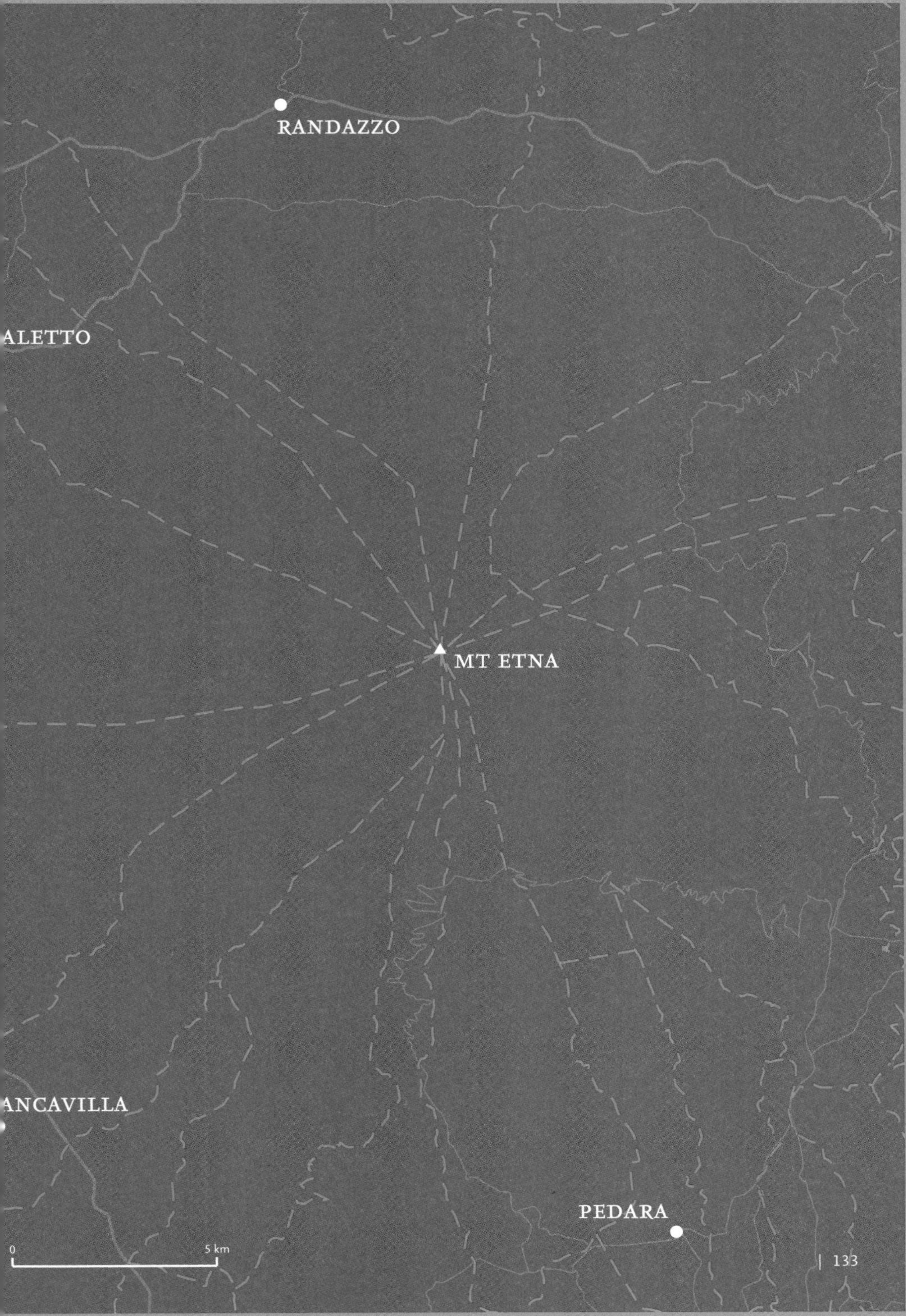

RANDAZZO
ALETTO
MT ETNA
ANCAVILLA
PEDARA
0
5 km

UNUSUAL INTERNAL BORDERS: ASIA

PUDUCHERRY

UNITED ARAB EMIRATES

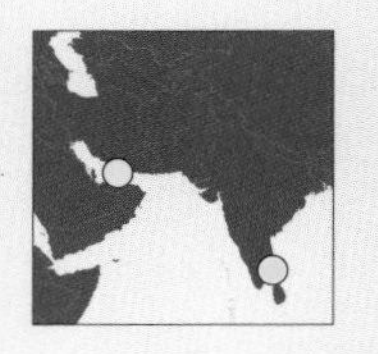

The small country with a large number of enclaves

Puducherry 11° 56' 17"N | 79° 48' 56"E
United Arab Emirates 24° 54' 25"N | 55° 35' 48"E

Jabal Hafeet Mountain, United Arab Emirates | Oman

Most countries in the world have their own internal borders, outlining the administrative areas into which they are divided. Just as with country borders, where the aim is to try and keep them as simple as possible, so that relations between the neighbouring countries are as good as they can be, internal borders should also be straightforward and logical. After all, it has been demonstrated many times throughout history that it is better for internal borders to be clearly defined, otherwise they can become a source of conflict (like in the examples of the disintegration of the USSR and Yugoslavia). Nevertheless, some internal borders around the world are rather strange.

Puducherry (currently the official name of Pondicherry) is an Indian union territory, one of the smallest – 492 square kilometres – yet at the same time also one of the most dispersed. This territory consists of four districts, each of which is an enclave. Some of these districts contain additional enclaves – the main one has eleven, known locally as pockets. Three districts (Puducherry, Yanam and Karaikal) are located in the Bay of Bengal, whereas the smallest district (Mahe, at nine square kilometres), is located on the Arabian Sea coast. Once again, colonial rule was the reason behind such fragmentation: these enclaves were French colonies, which were returned to India after the Second World War, and which, even though separated by some hundreds of kilometres, were constituted as a unified union territory.

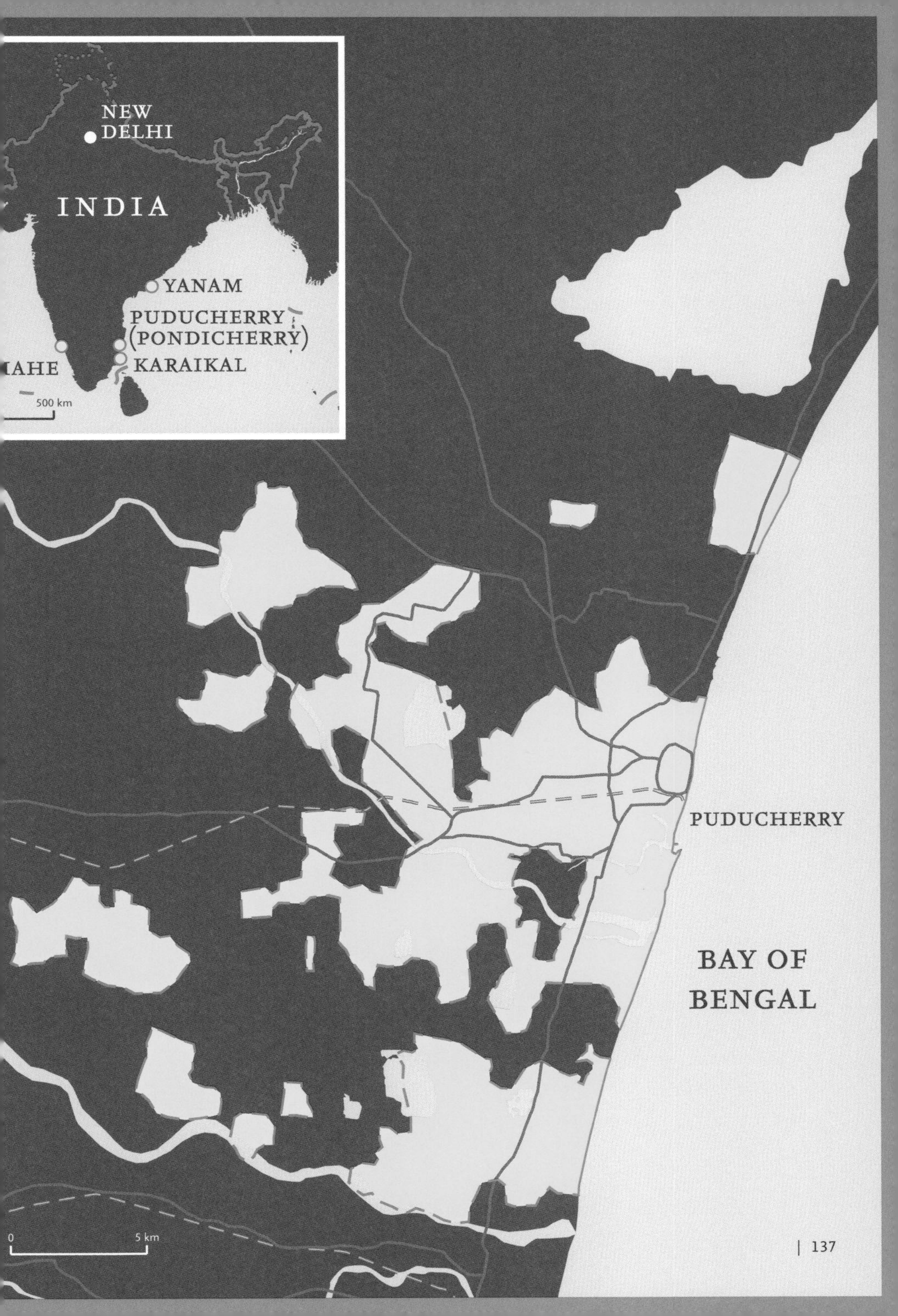
NEW DELHI
INDIA
YANAM
PUDUCHERRY (PONDICHERRY)
KARAIKAL
MAHE
500 km
PUDUCHERRY
BAY OF BENGAL
0
5 km

The situation is equally complicated with the internal borders of the **United Arab Emirates.** Except for the largest emirate, Abu Dhabi, all the other emirates consist of numerous enclaves and exclaves. The numbers on the map show the seven different Emirates.

The Republic of Indonesia, one of the largest and most densely populated countries in the world, consists of thirty-four provinces, two of which are distinctive, each in its own way. The westernmost province of **Aceh** is the only Indonesian province that fully enforces sharia law. On the other hand, in the southern part of the island of Java, is the Special Region of **Yogyakarta**, the only monarchy – a sultanate – among all the Indonesian provinces. The sultans of Yogyakarta acquired this right after their significant contribution to the fight for Indonesia's independence.

1 – Abū Z̧aby (Abu Dhabi)
2 – Dubayy (Dubai)
3 – Ash Shāriqah (Sharjah)
4 – 'Ajmān
5 – Umm al Qaywayn
6 – Ra's al Khaymah
7 – Al Fujayrah (Fujairah)

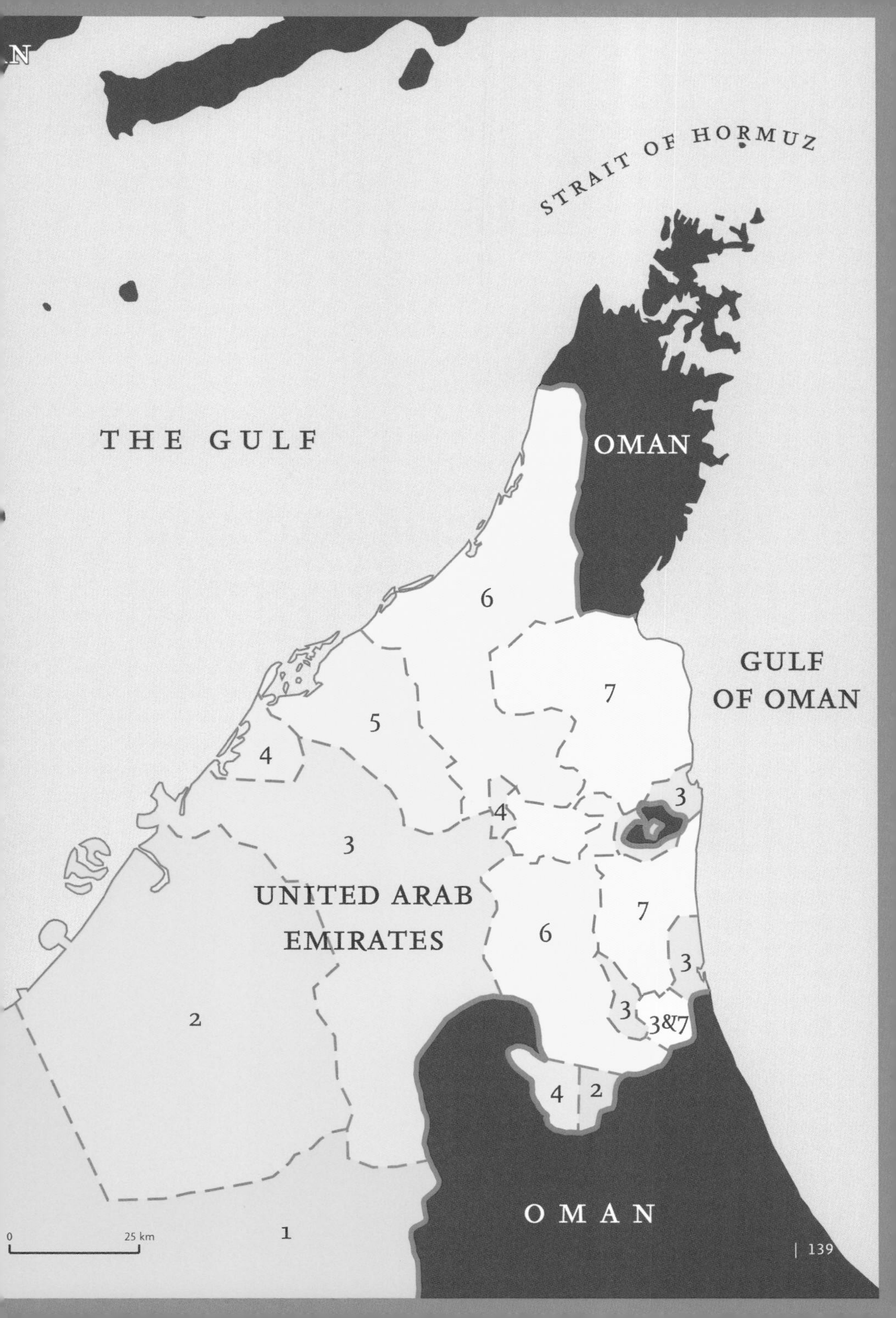
N
STRAIT OF HORMUZ
THE GULF
OMAN
GULF OF OMAN
6
7
5
4
3
4
3
UNITED ARAB EMIRATES
7
6
3
3
3&7
2
4
2
OMAN
1
0
25 km

UNUSUAL INTERNAL BORDERS: NORTH AMERICA

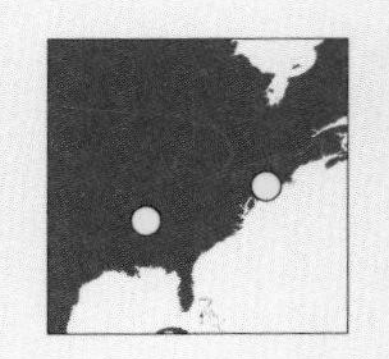

ELLIS ISLAND

KENTUCKY BEND

The land that will become an island, but only for a time

Ellis Island 40° 41' 57"N | 74° 02' 25"W
Kentucky Bend 36° 32' 45"N | 89° 30' 44"W

Ellis Island, USA

For millions of immigrants seeking their fortune in the USA between the end of the nineteenth and the middle of the twentieth centuries, **Ellis Island** was the place where they first disembarked. This island, less than a kilometre away from the Statue of Liberty and Liberty Island, covers an area of eleven hectares, but is divided by a rather unusual border between the federal states of New York and New Jersey. In the nineteenth century, New York and New Jersey reached an agreement that the island, which at the time was situated in New Jersey waters, should become an exclave of New York. As the island became too small for the requirements of the Immigration Service, it was greatly enlarged by land reclamation, its area being increased to approximately ten times its original size. New Jersey then pointed out that only the area of the natural part of the island belonged to New York, and that the artificial expansion belonged to New Jersey. Regardless of the borders of the federal states, Ellis Island has been the property of the United States government for the past two centuries.

The situation regarding the nearby **Liberty Island**, best known as the location of the Statue of Liberty, is also unusual. The island itself belongs to New York, but all the water surrounding the island, up to the coast itself, belongs to New Jersey, so that the Statue of Liberty is claimed by both American states.

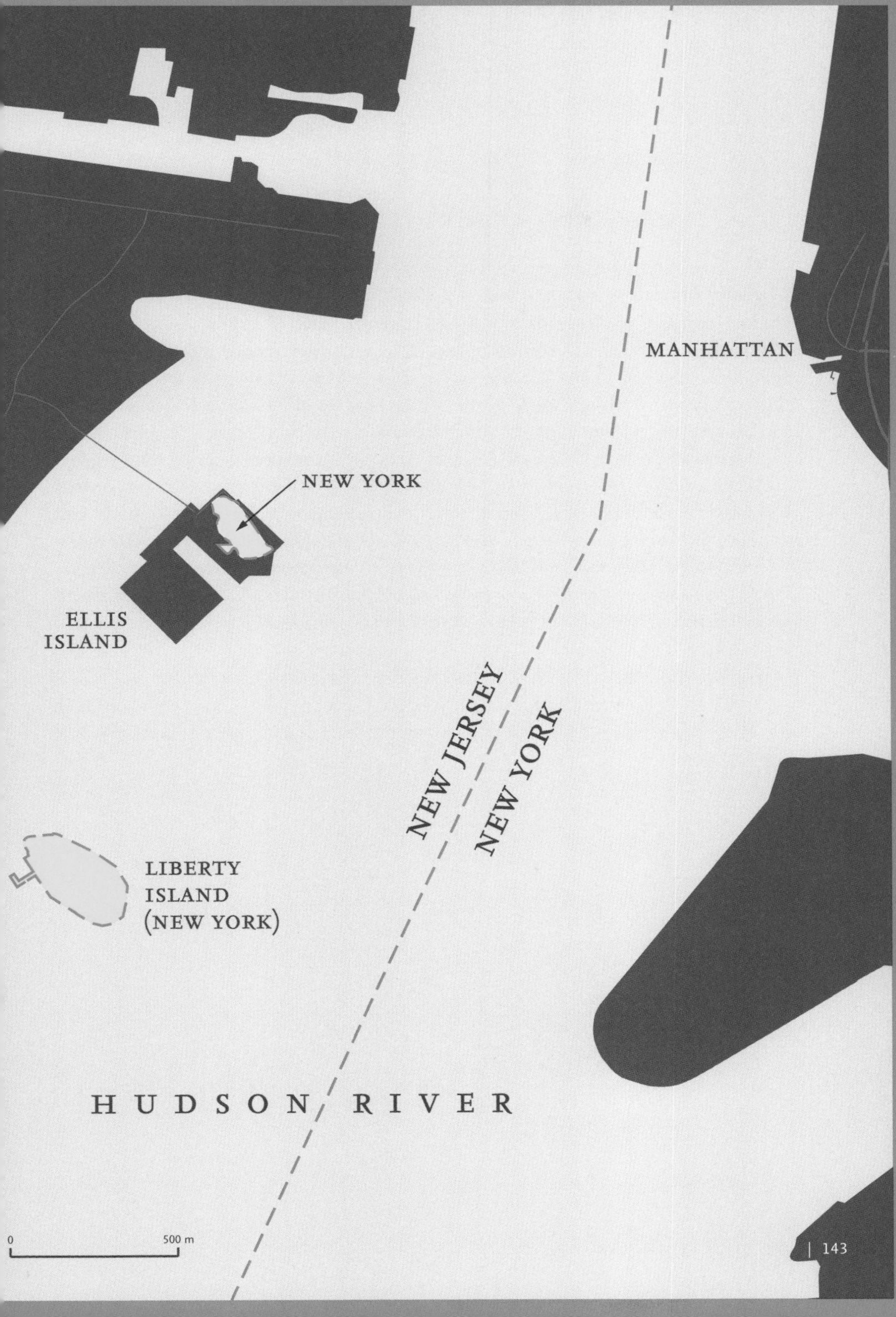

MANHATTAN

NEW YORK

ELLIS
ISLAND

NEW JERSEY

NEW YORK

LIBERTY
ISLAND
(NEW YORK)

HUDSON RIVER

0 500 m

The most southeasterly point of the American federal state of Kentucky is the **Kentucky Bend** peninsula – with less than twenty inhabitants – which was formed as the result of great meanders of the Mississippi River. The only land connection of the forty-five-square-kilometre peninsula is to the south, where it borders Tennessee. It is quite likely that the Mississippi River will fairly soon break through this connection with Tennessee, turning Kentucky Bend into an island, after which it will gradually connect with its northern neighbour, the state of Missouri.

The small US town of **Payne** (220 inhabitants) in the state of Georgia represented an enclave within the larger town of Macon (230,000 inhabitants). After numerous unsuccessful referendums, Payne was finally officially disbanded in 2015 and annexed to Macon. A similar situation also exists with the small towns of **Norridge** and **Harwood Heights**, which are officially enclaves within the city of Chicago. There are many such examples in the USA. In Pennsylvania alone, there are over 300 internal enclaves within cities, villages and municipalities.

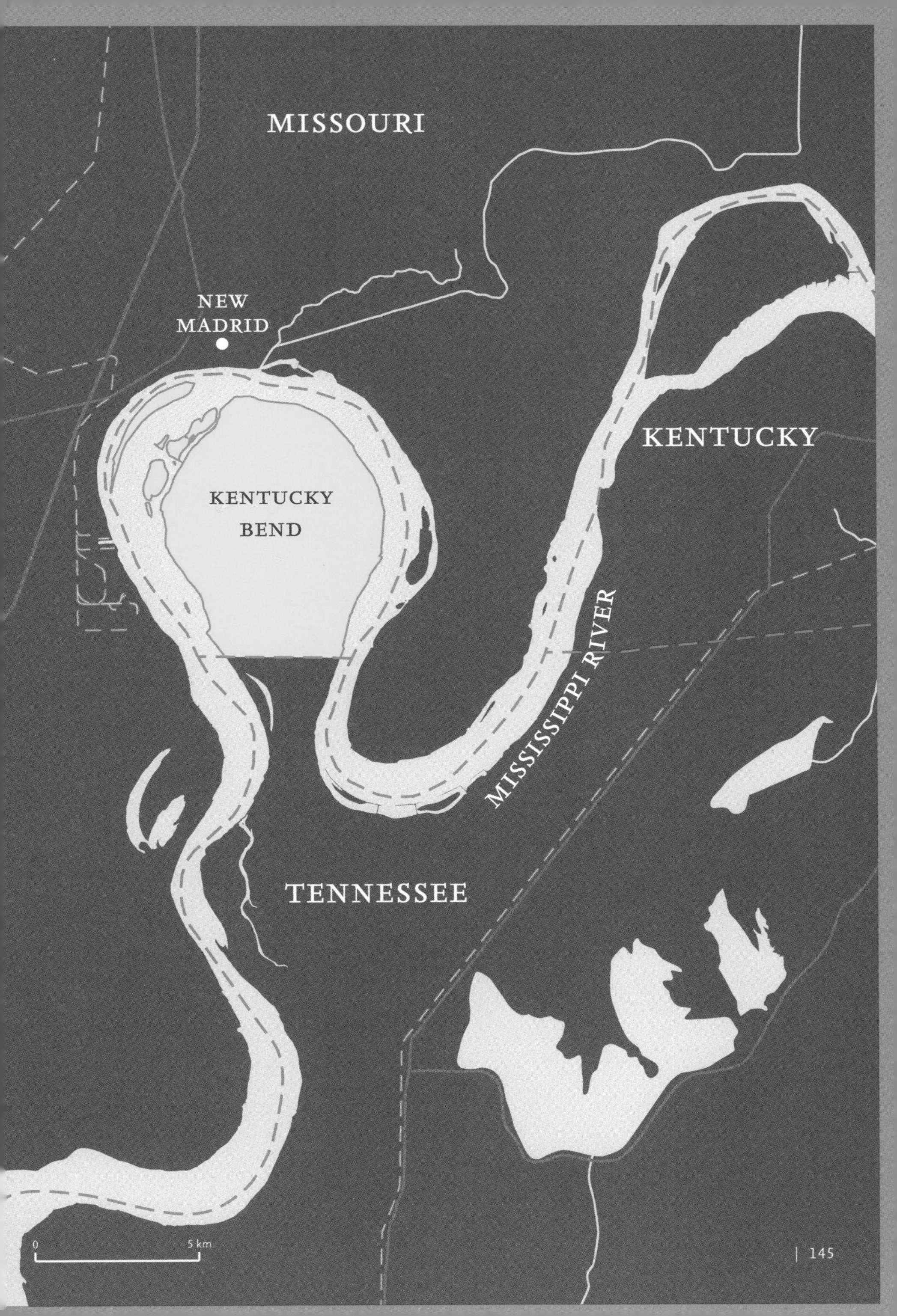
MISSOURI
NEW
MADRID
KENTUCKY
KENTUCKY
BEND
MISSISSIPPI RIVER
TENNESSEE
0
5 km

UNUSUAL INTERNAL BORDERS: EUROPE

BREMERHAVEN, HAMBURG
KOHTLA-JÄRVE
LIECHTENSTEIN
GĂGĂUZIA

The enclaves created to provide ports for cities

Bremerhaven 53° 32' 09"N | 8° 34' 58"E
Hamburg 53° 32' 58"N | 9° 59' 40"E
Kohtla-Järve 59° 23' 44"N | 27° 17' 00"E
Liechtenstein 47° 08' 47"N | 9° 33' 41"E
Găgăuzia 46° 16' 58"N | 28° 39' 41"E

Neuwerk, Germany

The smallest German federal state, officially the Free Hanseatic City of **Bremen**, consists of two enclaves covering a total area of 404 square kilometres, some sixty kilometres apart from each other and fully surrounded by the province of Lower Saxony. At the 1815 Vienna Congress, the city of Bremen was pronounced one of the thirty-nine sovereign states within the German confederation. Since the Weser River – on which Bremen stands, sixty kilometres from the mouth in the North Sea – deposited large quantities of sand, Bremen required a new port on the sea coast. The land was purchased from the Kingdom of Hanover of that time, and thus the new port was constructed there, named **Bremerhaven** (literally meaning 'the Port of Bremen'). Presently, this town is a second enclave which, together with the city of Bremen, forms the German federal state of Bremen. Additionally, the northeastern part of Bremerhaven, **Fehrmoor**, is detached from the rest of Bremerhaven by a narrow strip of land that belongs to Lower Saxony, making it in fact the third enclave forming this German province.

Another small German federal province, the Free and Hanseatic City of **Hamburg**, has its own sea enclave. The city-state of Hamburg is situated on the Elbe River about 110 kilometres south of its mouth in the North Sea. Since Hamburg believed that it would require space for the construction of a deep draft port, it was assigned two small islands in the North Sea, after the Second World War, not far from the mouth of the Elbe and some 120 kilometres away from Hamburg. The plan was for a big port to be constructed on the islands, but the enormous costs and ongoing rallies of environmental protesters resulted in the postponement then abandonment of the plans. Nowadays there are three islands in that enclave: two natural ones – populated **Neuwerk** (three square kilometres, forty inhabitants) and unpopulated **Scharhörn** – and an artificial one, **Nigehörn**, formed when it became apparent that the waves endangered the neighbouring Scharhörn and the bird sanctuary there. That entire enclave is now a national park, but also part of a whole series of coastal national parks along the North Sea coasts of Denmark, Germany and the Netherlands.

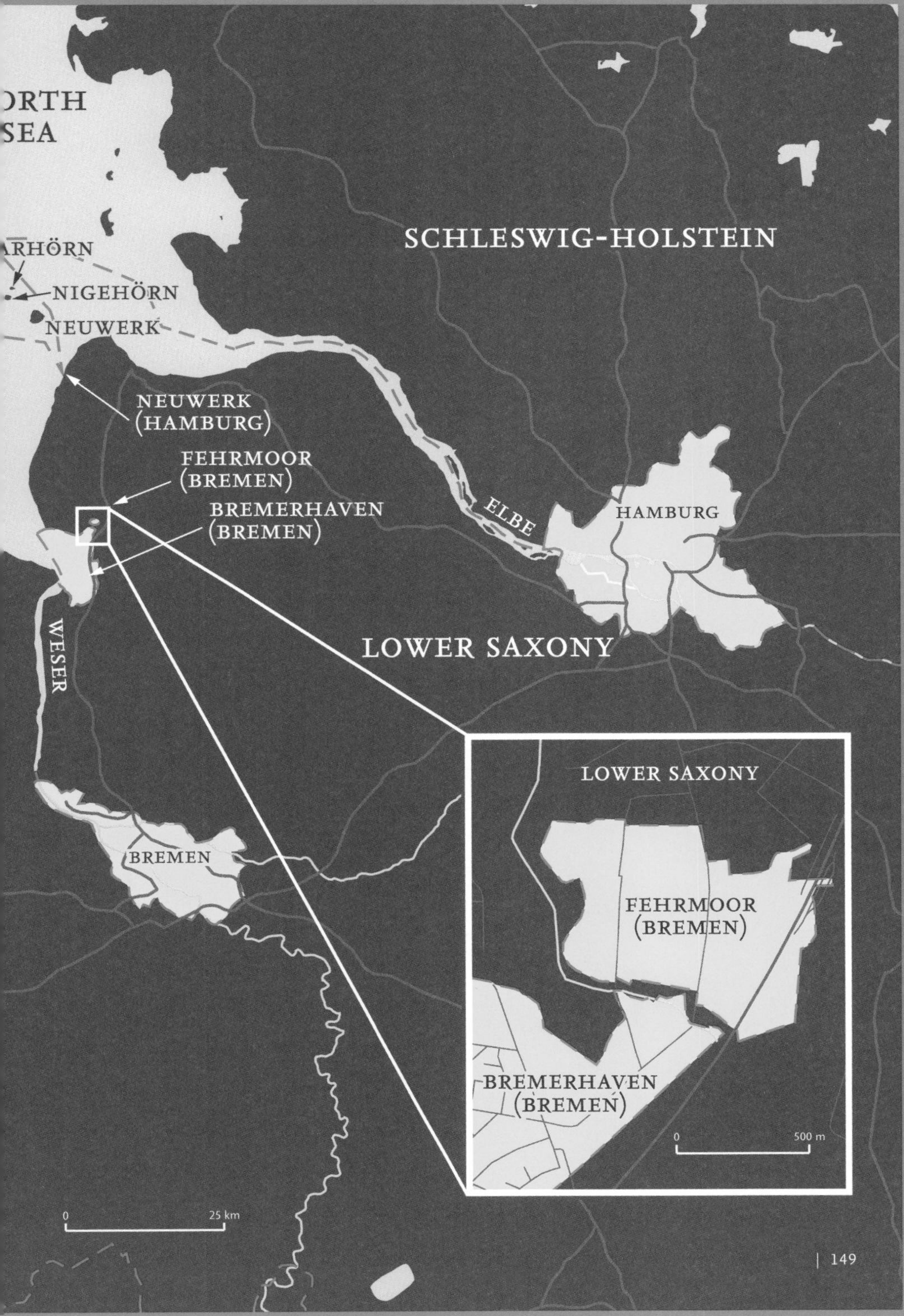

ORTH SEA

SCHLESWIG-HOLSTEIN

ARHÖRN

NIGEHÖRN

NEUWERK

NEUWERK (HAMBURG)

FEHRMOOR (BREMEN)

BREMERHAVEN (BREMEN)

ELBE

HAMBURG

WESER

LOWER SAXONY

BREMEN

0 25 km

LOWER SAXONY

FEHRMOOR (BREMEN)

BREMERHAVEN (BREMEN)

0 500 m

The territory of the Estonian city **Kohtla-Järve**, consisting of five enclaves (Ahtme, Järve, Kukruse, Oru and Sompa), has also been unusually broken into separate parts. The town has about 37,000 inhabitants – the majority being of Russian ethnicity – and the distance between the two furthest enclaves is about thirty kilometres. During the time of the USSR, several more enclaves belonged to this city, and these are now separate towns. Two further enclaves (Viivikonna and Sirgala) were taken out of the municipality after the administrative reform of 2017.

Although one of the smallest countries in the world, the Principality of **Liechtenstein** has rather jagged internal borders with a large number of enclaves.

Certain cantons in **Switzerland** also consist of several non-contiguous segments, such as Schaffhausen (in which the German exclave Büsingen is situated), Solothurn and Fribourg/Freiburg.

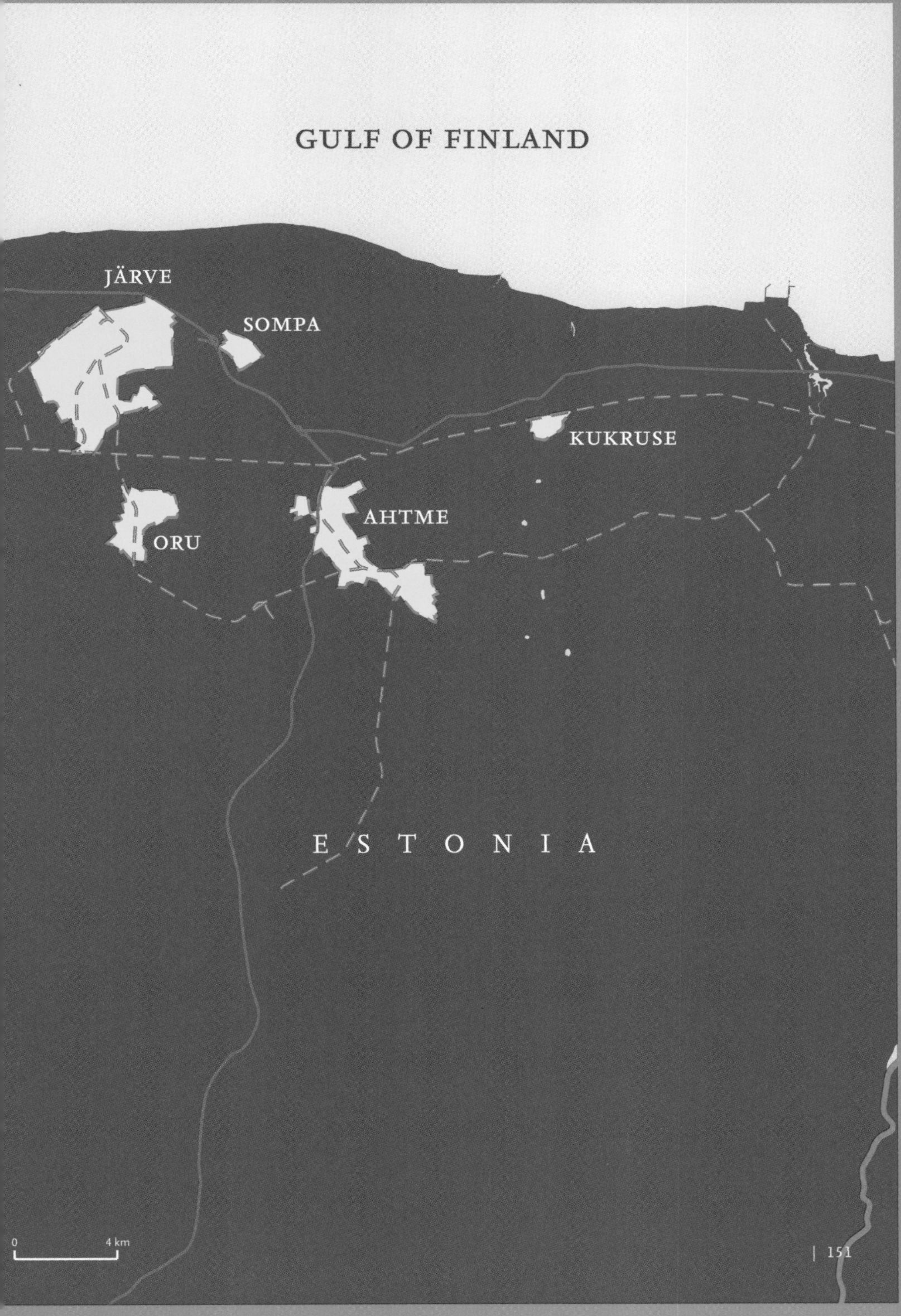

GULF OF FINLAND

JÄRVE

SOMPA

KUKRUSE

AHTME

ORU

ESTONIA

0 4 km

Găgăuzia, is an autonomous region in Moldova. This region was named after the majority population, the Găgăuz people, who probably originate from the Seljuq Turks, though they converted to Eastern Orthodox Christianity back in the eighteenth century. At the beginning of the nineteenth century they populated the region of the Russian Bessarabia, where they still live. After the declaration of Moldovan independence in 1991, it took several years of negotiations before the Găgăuzian autonomy was proclaimed. A ruling was passed that all places with more than 50 per cent of Găgăuz, as well as also those with 40–50 per cent that decided by referendum, should become part of the autonomy, and thus it resulted in Găgăuzia consisting of four enclaves in the southern part of Moldova. According to the constitution of Moldova, in the event that the country should change its status (such as becoming united with Romania, which is what many Moldovans want), Găgăuzia would have the right to proclaim its independence.

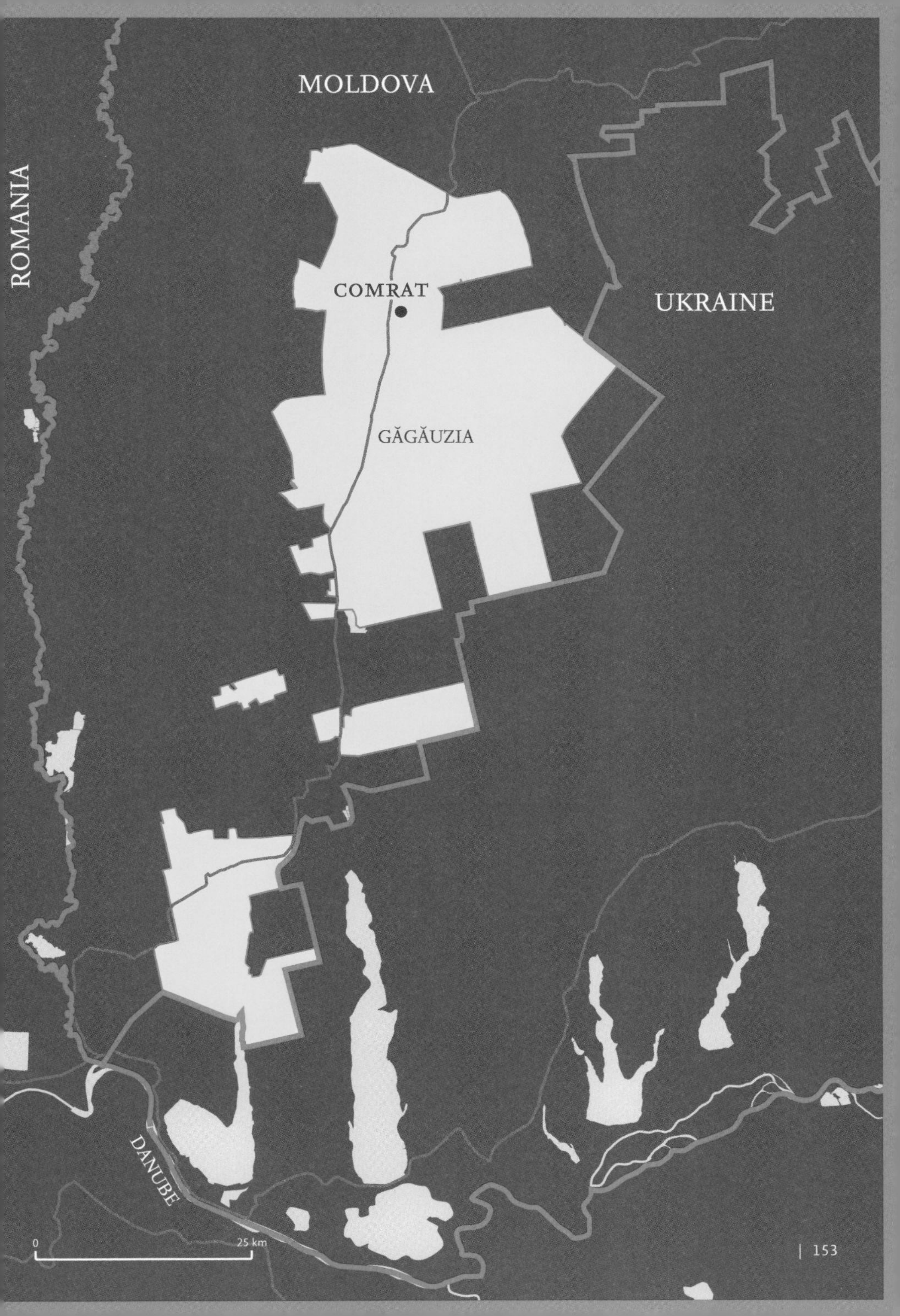
MOLDOVA
ROMANIA
COMRAT
UKRAINE
GĂGĂUZIA
DANUBE
0
25 km

UNUSUAL INTERNAL BORDERS: AUSTRALIA

JERVIS BAY TERRITORY

BOUNDARY ISLET

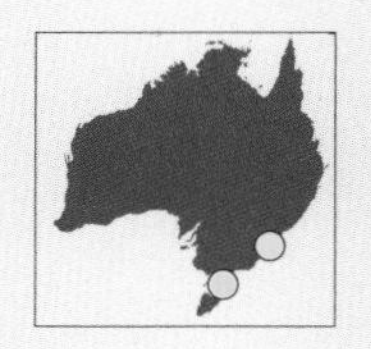

The state border only 85 metres long

Jervis Bay Territory 35° 09' 20"S | 150° 41' 25"E
Boundary Islet 39° 11' 55"S | 147° 01' 17"E

Point Perpendicular Lighthouse, Jervis Bay, Australia

Although Australia does not have land borders with any other country, that does not mean that the smallest continent (and the biggest country in it) does not have any unusual interior borders.

Australia is a federation, consisting of six states (Western Australia, New South Wales, Queensland, South Australia, Tasmania and Victoria), two major mainland territories (Northern Territory and Australian Capital Territory) and several smaller territories.

The first oddity relates to the territory of the Australian capital (namely ACT – **Australian Capital Territory**). This territory is an enclave within the state of New South Wales, and is about a hundred kilometres away from the Tasmanian Sea. When ACT was established, a decision was passed that this region must have access to the sea. In order to realize this, **Jervis Bay Territory** was established on the coast, in the southern part of the homonymous bay (about 150 kilometres from Canberra). This territory de facto belonged to ACT until 1989 – it was in that year that ACT acquired broader interior powers, and Jervis Bay Territory became yet another territory within the Australian federation. However, since the Australian constitution stipulates that ACT should have sea access, it was allegedly decided that the northern part of Jervis Bay, the **Beecroft Peninsula**, 130 kilometres south of Sydney, would pertain to ACT as an exclave. It is interesting that the very tip of this peninsula, with the lighthouse, remained a territory of New South Wales, so that now this ACT exclave surrounds the exclave of New South Wales. The Beecroft Peninsula is mostly unpopulated and is used for Australian army training. On the other hand, according to some information, the southern part of Beecroft Peninsula actually belongs to the federal government, but is still within New South Wales. What is true is uncertain at this time.

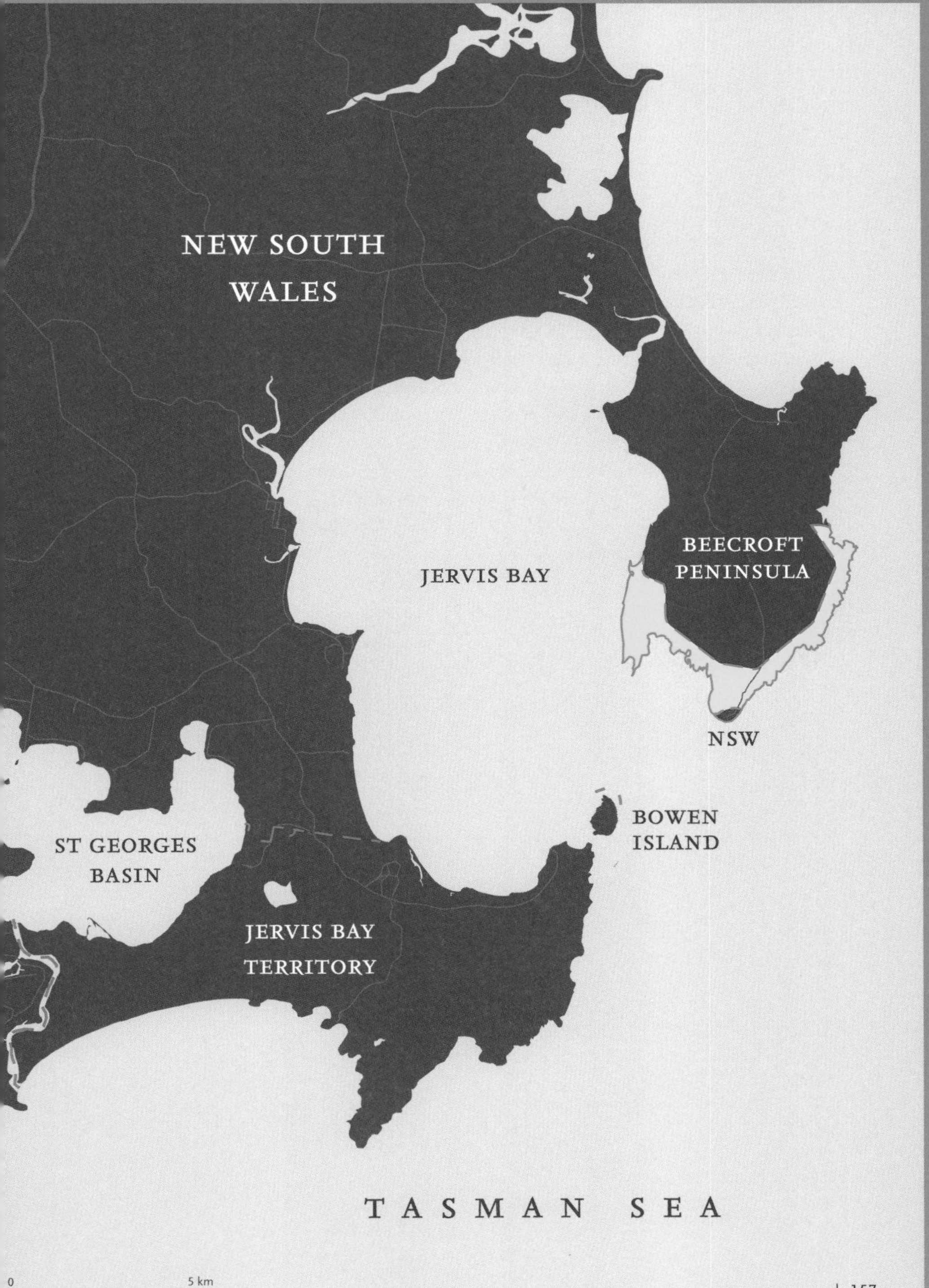

NEW SOUTH WALES

JERVIS BAY

BEECROFT PENINSULA

NSW

BOWEN ISLAND

ST GEORGES BASIN

JERVIS BAY TERRITORY

TASMAN SEA

0 5 km

Another oddity has to do with the land border of the Australian states of Victoria and Tasmania. As Tasmania is an island state, the land border of these states is not shown on any map. However, it does exist. It is situated on the tiny **Boundary Islet**, in Bass Strait, which separates Tasmania from Australia. When the northern sea border of Tasmania was defined, it was decided that the line would be drawn along the latitude of 39° 12'. It was only later that it was established that Boundary Islet did not lie further north, as was believed until then, but that it was situated exactly at that latitude, and thus it was given that name and the shortest boundary between the Australian states, only eighty-five metres long. The islet covers an area of about six hectares, and represents a bare rock, whipped mercilessly by wind and waves.

VICTORIA

TASMANIA

BOUNDARY
ISLET

HOGAN
ISLAND

BASS STRAIT

0 1 km

UNUSUAL CAPITALS

The only nation whose official capital is a ghost town

Plymouth, Montserrat

Probably everyone knows that the capital city of a country is the city in which the president/monarch, government, parliament and supreme court are situated. However, is the situation in every country so simple? Of course not.

For instance, the official capital of **Estonia** is Tallinn, although the Supreme Court and certain ministries are located in Tartu, which has also been the university centre of Estonia since the middle of the seventeenth century.

The capital of **Germany** is Berlin, with the seat of the chancellor and president. However, Bonn is the seat to several ministries, as well as the second seat of the chancellor and president. The city of Karlsruhe is the seat of most judicial authorities, including the Federal Constitutional Court.

The capital of **Montenegro** is Podgorica, but the official seat of the president is the former royal capital, Cetinje.

South Africa does not state its capital in its constitution, but specifies that Pretoria is its administrative centre, as well as the seat of the president and government (most embassies are located in Pretoria); Cape Town is the legislative centre, as well as the seat of parliament; Bloemfontein is the judiciary seat, as the Supreme Court is located there. As if this situation was not complicated enough, the Constitutional Court is located in Johannesburg.

In contrast, there are countries that, for various reasons, do not have an official capital city at all. This applies, for example, to **France, Portugal** and **Switzerland**, whose constitutions do not define any city as the capital, although of course, almost all state bodies are located in Paris, Lisbon and Bern respectively. Some countries are too small, so they do not have a capital – these include the **Vatican City, Singapore** and **Monaco**, which are all city-states. A small country in the Pacific Ocean, **Nauru**, does not officially have a capital, although Yaren, the largest settlement on the island, is the seat of its government.

In some countries, the capital is not the seat of government. Such is the case in **Georgia**, whose capital, Tbilisi, is the seat of the president and the Supreme Court, but Kutaisi, in the west, is the seat of both parliament and government. According to the constitution of the **Netherlands**, Amsterdam is the capital, although the government, parliament, Supreme Court and royal castle are located in The Hague, as well as most of the embassies. Sucre is officially still the capital of **Bolivia**, although practically all state bodies are found in La Paz. It is interesting that La Paz is the world's highest capital (3,640 metres above sea level), while Sucre is also high on this list, since at 2,750 metres it is the third highest. The situation in **Chile** is interesting: the capital of this extremely elongated country is Santiago de Chile, but the seat of the National Congress (parliament) is located some fifty kilometres away in the coastal city of Valparaíso (whose Spanish name means 'paradise valley'). In 1983, **Côte d'Ivoire (Ivory Coast)** officially declared Yamoussoukro, a city in the central part of the country, as its capital but almost all state bodies and embassies are still located in coastal Abidjan, a megapolis with almost five million inhabitants. Many countries in the world have new capitals built outside the territory of the existing capital: **South Korea** has recently transferred a large number of its ministries and agencies to Sejong City, **Malaysia** moved the seat of its government to Putrajaya, while **Myanmar (Burma)** transferred its capital to Naypyidaw, some 320 kilometres north of Yangoon.

It is interesting to note that there are constituent divisions of some countries that have several capital cities. This is the case for the **Azores**, an autonomous region of Portugal, which has three capital cities, each on its 'own' island – Ponta Delgada is the seat of its autonomous government, the parliament is in Horta, while Angra do Heroísmo is the judicial centre and historically the capital of the island group. The **Canary Islands**, a Spanish autonomous community not far from the coast of Western Sahara and Morocco, has two equal capitals: Santa Cruz de Tenerife and Las Palmas de Gran Canaria. The official capital of the **Republika Srpska**, one of the two Bosnian political entities, is Sarajevo, even though almost all governing bodies are in Banja Luka. **Canton 10** (or the Herzeg-Bosnia County) has three capitals: Kupres is the seat of the prefect, Tomislavgrad is the legislative centre, while the government of this canton of the Federation of Bosnia and Herzegovina is in Livno. The Caribbean island of **Montserrat**, a British Overseas Territory, currently may have the strangest situation in terms of its capital. A volcanic eruption destroyed half of the island in 1995, including the capital, Plymouth. The temporary headquarters of the local government is currently in the northern town of Brades, while a new capital is being built in Little Bay. However, deserted Plymouth is technically still the capital, making it the only ghost town that is officially the capital of a country or territory.

CHANDIGARH

INDIA

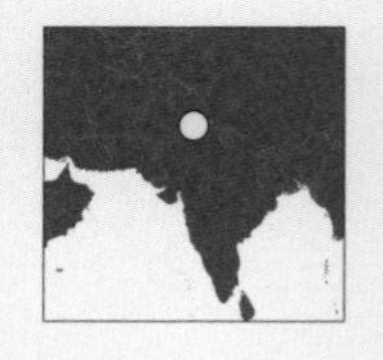

The city that is capital of two states and a union territory

30º 43' 51"N | 76º 45' 41"E

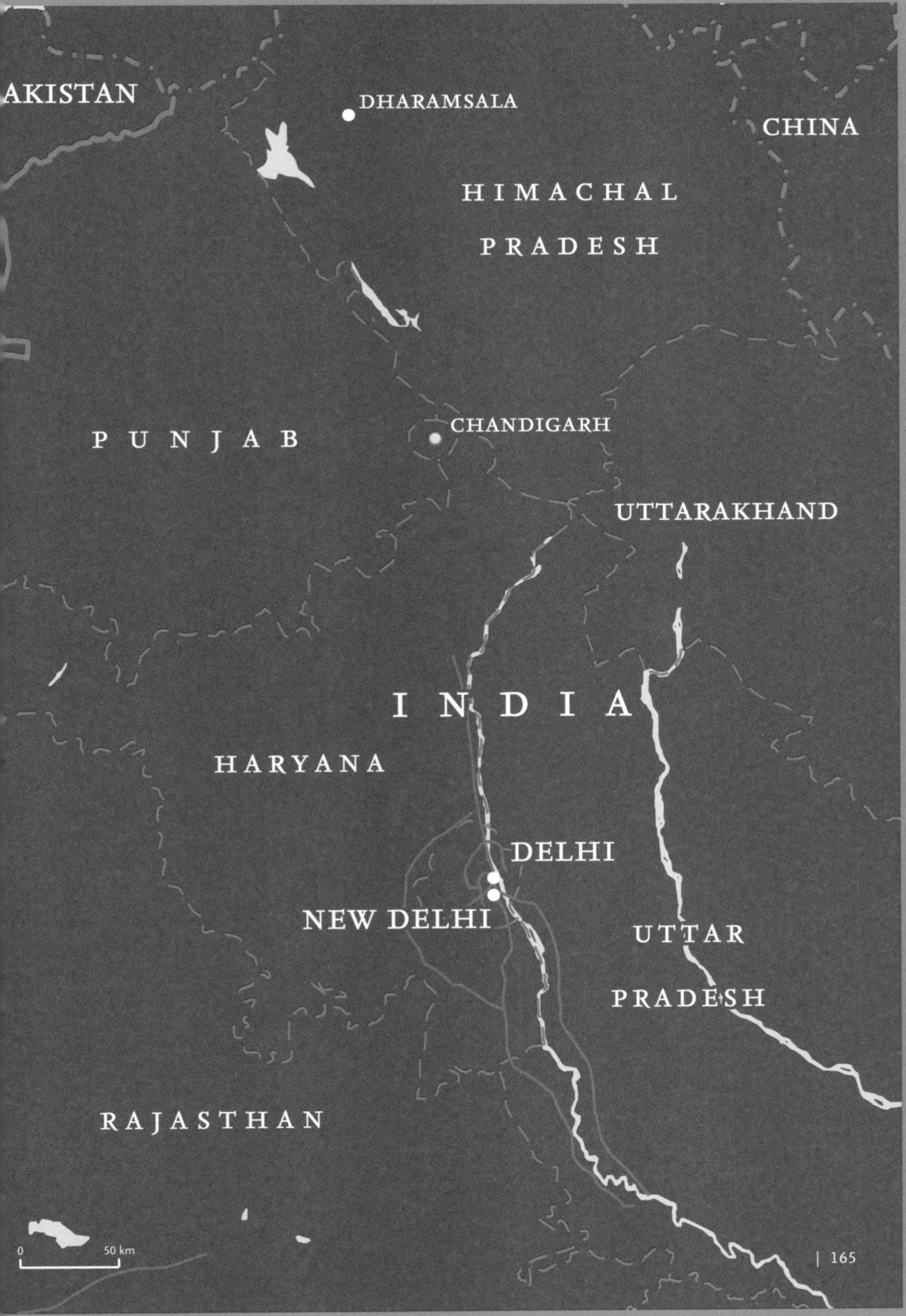

AKISTAN

DHARAMSALA

CHINA

HIMACHAL PRADESH

PUNJAB

CHANDIGARH

UTTARAKHAND

INDIA

HARYANA

DELHI

NEW DELHI

UTTAR PRADESH

RAJASTHAN

0 50 km

In many countries around the world, the capital city occupies its own administrative unit, which is separate from the rest of the country. This situation exists in the USA (**Washington D.C.** is a separate district, leased to the US federal government by the states of Maryland and Virginia), Argentina (the capital is officially called the Autonomous City of **Buenos Aires**), Australia (**Canberra** does not belong to any of the Australian federal states, but to the Australian Capital Territory, which is an enclave within New South Wales), Austria (the city of **Vienna** is the capital of Austria but also one of the federal states) and Ethiopia (**Addis Ababa** has a similar status to Vienna).

However, in rare cases, the capital of a constituent area may also be separated from the area of which it is the capital. One example of this is the city of **Chandigarh** in India. This city serves as the capital of two Indian states, Punjab and Haryana, but officially it is not a part of either of them, but has the status of an Indian union territory.

In order to understand this unusual situation, we must go back a little in history. After the Second World War, India gained independence from the United Kingdom, but at the same time it was divided between the predominantly Muslim Pakistan (which also included the present-day Bangladesh) and secular India. During this division, a part of Punjab, with its capital Lahore, was conferred to Pakistan, while the remaining part formed the Indian state of Punjab. At the initiative of the well-known prime minister of India, Nehru, a new capital was built – Chandigarh – for the Indian Punjab. Only a decade later, the southern part of Punjab broke away from Punjab (in which Punjabi was mainly spoken) and formed the state of Haryana, with Hindi as the prevalent language. As Chandigarh was right on the border of these states, it was decided that it would remain a 'double' capital. This solution was supposed to be only temporary, until Haryana had built its own capital. However, this has still not happened.

Today, Chandigarh has the status of an Indian union territory, while at the same time it is the capital of both neighbouring states. The oddity is reflected in the fact that Chandigarh has almost the same status in the Indian federation as the states it serves as a capital. In addition, the city of Chandigarh is classed as the capital of the union territory of Chandigarh.

This city is well known for its modern architecture and urban style, which is not surprising given the fact that the general plan was designed by the famous Swiss-French architect, Le Corbusier. Apart from its architecture, Chandigarh can be proud of its lush greenery, general cleanliness and one of the highest per capita incomes in the whole of India.

Recent years have seen the development of Greater Chandigarh, which includes the territory of Chandigarh and several surrounding cities in neighbouring Punjab, Haryana and Himachal Pradesh (the latter being the state in which Dharamshala is located, the seat of the Tibetan government in exile).

Another city, several hundred kilometres to the south, is also a capital of two Indian states. It is the city of **Hyderabad**, a long-standing capital of the state of Andhra Pradesh. When the northern part broke off to form a new federal state, Telangana, Hyderabad remained within its territory. According to an agreement at that time, Hyderabad would remain a joint capital for a maximum of ten years, until Andhra Pradesh had finished building its new capital, Amaravati.

Note: India consists of union territories and states; the main difference between them is that the states elect their presidents democratically, while the president of India appoints administrators to the territories.

Assembly Building, Chandigarh, India

TORNIO AND HAPARANDA

FINLAND | SWEDEN

The merged city that celebrates the New Year twice

65° 51' 06"N | 24° 08' 38"E

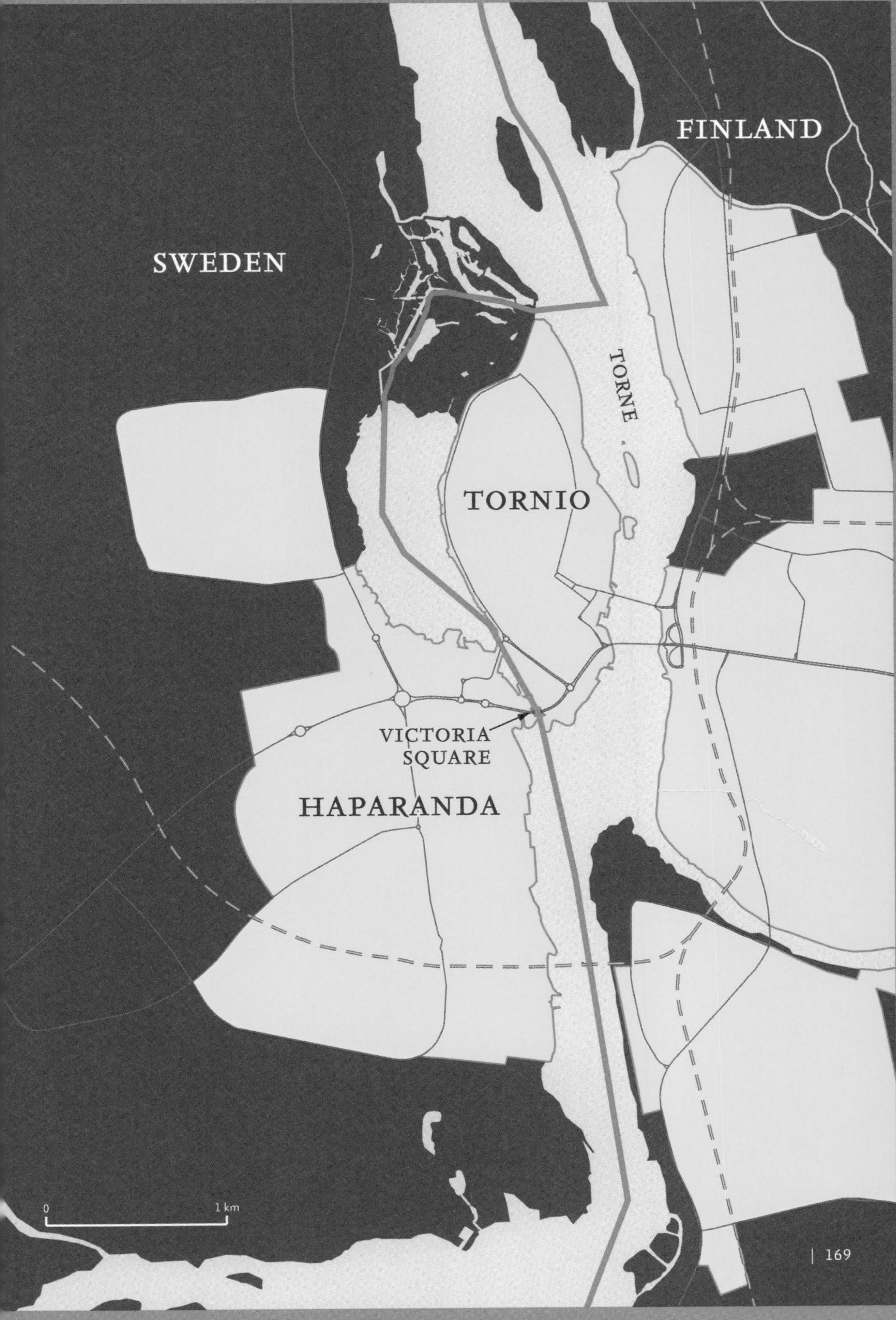

FINLAND

SWEDEN

TORNE

TORNIO

VICTORIA
SQUARE

HAPARANDA

0 1 km

Torne or Tornio is a river that originates in northern Sweden, and which, from about halfway along its course to its mouth in the Gulf of Bothnia, represents the border between Sweden and Finland. The cities **Tornio** (Finland) and **Haparanda** (Sweden) are situated at the mouth of the river. As the border between these countries has essentially been completely open for the past fifty years, this has virtually led to the merging of these cities. This has been helped by the entry of Sweden and Finland into the European Union, as well as the later signing of the Schengen Agreement by both countries.

The city of Tornio was founded by the Swedes in the seventeenth century, on a large island at the mouth of the river with which it shares its name. At that time, Finland belonged to Sweden and Tornio was an important trading town. The residents of the city mostly spoke Swedish, though the Finnish language prevailed in the villages around the city. The Sami people were living north of the city. The city developed until the early nineteenth century, when, following the Swedish–Russian war, the whole of Finland (including the city of Tornio) became a part of Russia. The city of Tornio then acquired the insignificant role of the far northern garrison, and the economic decline that followed continued for a long time, even when Finland gained independence after the Second World War. This was the reason why most Swedes left the city and, on the opposite (Swedish) bank of the Tornio river, raised a new city named Haparanda.

After the Second World War, both cities developed rapidly, and the border between the two Scandinavian neighbours became less significant. Over time, this has led to a kind of unification of the Finnish Tornio and Swedish Haparanda. Today, the two cities are considered twin cities, and the city councils are working on a plan to completely merge them to form a single city, under the combined name of either TornioHaparanda or HaparandaTornio. In terms of this plan, much has already been completed, and many city services are already integrated, while most shops on either side of the river accept both euros and Swedish kronor. A new central square of the unified city is being built right on the border line and is named Victoria Square, after Sweden's crown princess.

An interesting fact is that Sweden and Finland are in different time zones. A consequence of this is the 'double' New Year's Eve celebration that takes place on the streets and squares of 'TornioHaparanda', which has become a tourist attraction. A great golf course is another attraction; it's position on the border makes it possible to hit the ball in one hour and for it to hit the ground in the previous hour.

LA CURE

FRANCE | SWITZERLAND

The cross-border hotel that welcomed German soldiers in one half while members of the French Resistance stayed in the other

46° 27' 54"N | 06° 04' 24"E

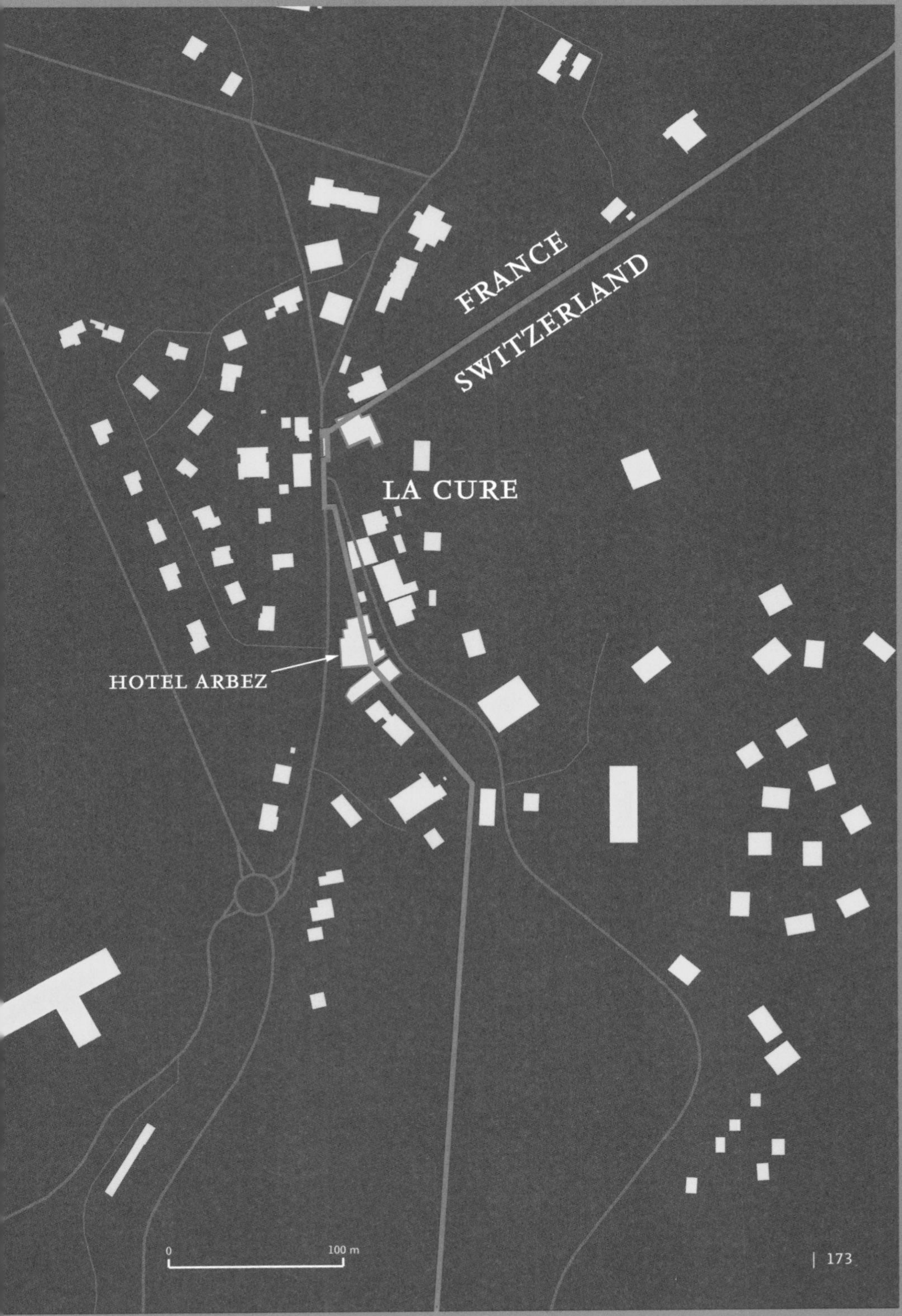

FRANCE

SWITZERLAND

LA CURE

HOTEL ARBEZ

0

100 m

La Cure is a small village on the border of France and Switzerland, about thirty kilometres north of Geneva. In this case, 'on the border' may be taken literally, since part of the village lies in Switzerland, and part in France. The border divides the village, streets, and even some buildings, the most famous being Hotel Arbez.

Until the second half of the nineteenth century, La Cure was completely in France. When France and Switzerland were establishing the exact position of the rather complex border at that time, the two countries exchanged some small pieces of territory and the border ended up being drawn directly through La Cure. According to a treaty, it was agreed that existing buildings were not to be disturbed, even if the boundary bisected them.

The most well-known building in La Cure is Hotel Arbez, where the French–Swiss border bisects several rooms. The border even passes through the honeymoon suite, dividing the double bed, which presents a tourist attraction.

The hotel was constructed in an interesting way. The treaty signed by the French and Swiss authorities anticipated that any buildings existing at the time of its ratification would not be torn down. An enterprising young man bought some land on the site of the future border and quickly constructed a building before the treaty became valid. After the ratification, he opened a grocer's shop on the Swiss side of the building and a bar on the French side. The sons of this man later sold the building to the grandfather of today's owner, who turned the building into a hotel and restaurant.

During the German occupation of France in the Second World War, the German soldiers were allowed to stay in the French (occupied) part of the hotel, while crossing to the Swiss side was strictly forbidden. It was not unusual for German soldiers to dine in the French restaurant, while members of the French Resistance movement were staying in the rooms on the Swiss side.

VALGA AND VALKA

ESTONIA | LATVIA

Where residents needed their passports when they visited neighbours across the street

57º 46' 28"N | 26º 01' 25"E

VALKA
VALGA
ESTONIA
LATVIA
1 km

Borders may exist for some time, then they may be moved, abolished, and new ones established. Occasionally, because of such changes, new borders may be drawn through unexpected places, for example through a town.

The town of **Walk** was founded in the Middle Ages by German knights who ruled the area belonging to today's Baltic republics of the former USSR. For almost two centuries this town was even the seat of parliament (Landtag) of the Livonian Confederation, one of the most important states of German chivalric orders. After the First World War, and the independence of Estonia and then Latvia, it was uncertain which country should claim Walk. With the aid of an international commission, it was decided that Estonia should get most of the town and Latvia a smaller part. Estonia named its part **Valga**, and Latvia, **Valka**. It was in Valka that the independence of Latvia was proclaimed for the first time and its red-white-red flag raised. After the Second World War, Estonia and Latvia were incorporated, not quite voluntarily, into the USSR, so the former international border became the border of the Soviet republics. This lasted until the 1990s, when Estonia and Latvia gained independence again. The border was once again established in the middle of the town, new barbed wires were emplaced, and the residents needed passports to visit their neighbours across the street. Fortunately, this situation lasted only until the beginning of the twenty-first century, when both countries entered into the European Union, and a little later the Schengen Area.

From then on, the development of Valga/Valka began to gather pace: boundaries were lifted, border crossings and fences removed, and the two towns gradually merged. Complete integration of many of the city services, such as transport, is planned. The introduction of the euro has also made life easier for the inhabitants of the town, whose motto is 'One town, two countries'. Valga today has a population of 13,000 inhabitants and an area of about sixteen square kilometres, while Valka is somewhat smaller, with about 6,000 people and an area of fourteen square kilometres.

FRANKFURT (ODER) AND SŁUBICE

GERMANY | POLAND

The town split in two when the country border moved

52° 20' 54"N | 14° 33' 14"E

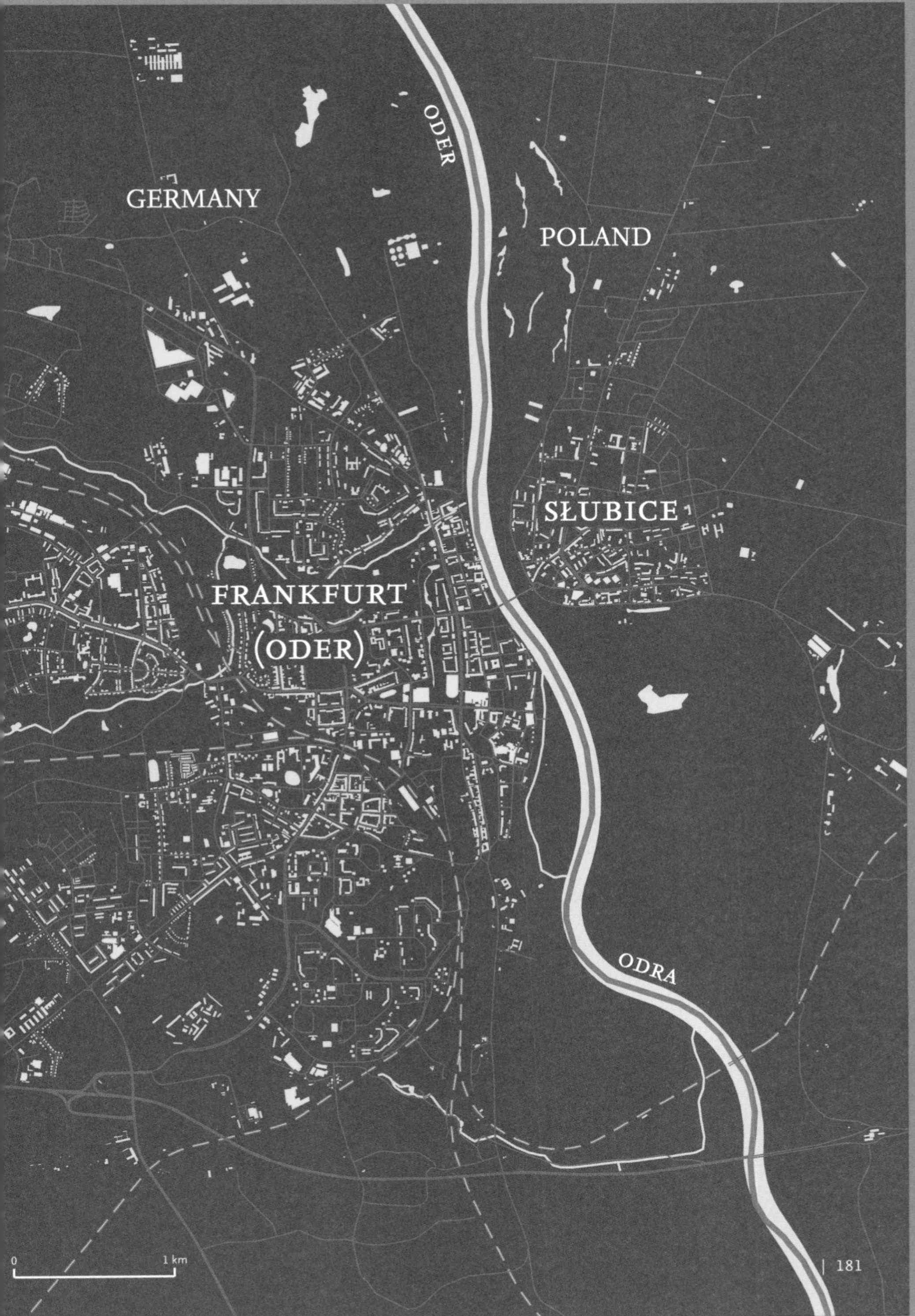

ODER

GERMANY

POLAND

SŁUBICE

FRANKFURT
(ODER)

ODRA

0 1 km

After every major war, there is peace. The treaties that establish this peace will often call for borders to be redrawn, imposed by those who won the war. This was the case after the Second World War, when Poland gained significant territory from Germany by moving the border all the way to the Oder (Odra) river. Suddenly, what had been an 'internal' river became a border between two countries, and the town of Frankfurt (Oder) a namesake of a bigger and better-known Frankfurt am Main, was split into two.

Frankfurt (Oder) is in the German federal state of Brandenburg. The population of this town has been decreasing for quite some time now (from almost 90,000 in the 1980s to about 60,000 thirty years later). It was officially given town status in the mid-thirteenth century, although it has been inhabited since much earlier. The first settlement was probably on the left bank, with the settlement on the right bank – now Polish – appearing later. In the nineteenth century Frankfurt (Oder) was one of the largest economic centres of the Kingdom of Prussia and the German Empire (Prussia being part of the empire), with the second largest annual fair in the entire empire – only the Leipzig Fair was larger.

During the Second World War, there were not many major battles over this Frankfurt, though that does not mean there was no destruction: the Soviet Red Army burnt a completely abandoned and empty town on its way to invade Berlin. After the war, the border between the two new communist neighbours, East Germany and Poland, was drawn along the Oder river, so that the eastern part of Frankfurt, Dammvorstadt, was politically separated from the rest of the town. Shortly afterwards, it was renamed Słubice.

Słubice is the Polish town opposite Frankfurt, on the right bank of the Oder. The town has a population of almost 20,000, and is named after the former nearby West Slavic settlement Zliwitz, which is mentioned in Frankfurt's city charter from the mid-thirteenth century. At that time, the Brandenburg Grafs (counts) bought land from Boleslaus II, the Duke of Silesia. From then on until the Second World War, Frankfurt (Oder) developed as a unitary town.

Presently, Frankfurt and Słubice are tightly connected towns. When Poland entered the European Union in 2004 and joined the Schengen Agreement in 2007, the border between the two parts of the former unitary Frankfurt was practically abolished. Various town services are gradually being merged and the towns are cooperating on various projects, such as the joint water treatment plant in Słubice. The Viadrina European University, which was founded in the sixteenth century and re-established at the end of the twentieth century, is located in Frankfurt. Together with the Adam Mickiewicz University from the Polish city of Poznań, Collegium Polonicum was founded in Słubice, making the combined town of Frankfurt/Słubice a significant European university and science centre.

Many children from Słubice go to preschools in Frankfurt, while more than 2,000 Poles live in the German town, and several hundred Germans live in Słubice. City public transport already has lines covering both towns, while tour guides show cultural and historical sites on both sides of the river; these sites are being jointly renovated. The Polish language is increasingly being taught in German preschools and schools, and vice versa, the German language in Polish ones.

GORIZIA AND NOVA GORICA

ITALY | SLOVENIA

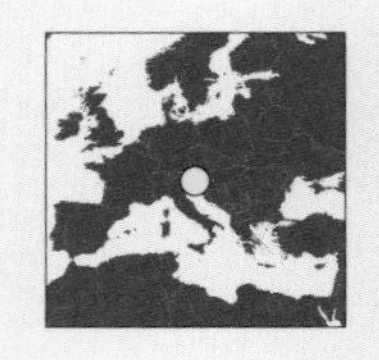

Even the names on the Slovene gravestones were Italianized

45° 57' 11"N | 13° 38' 00"E

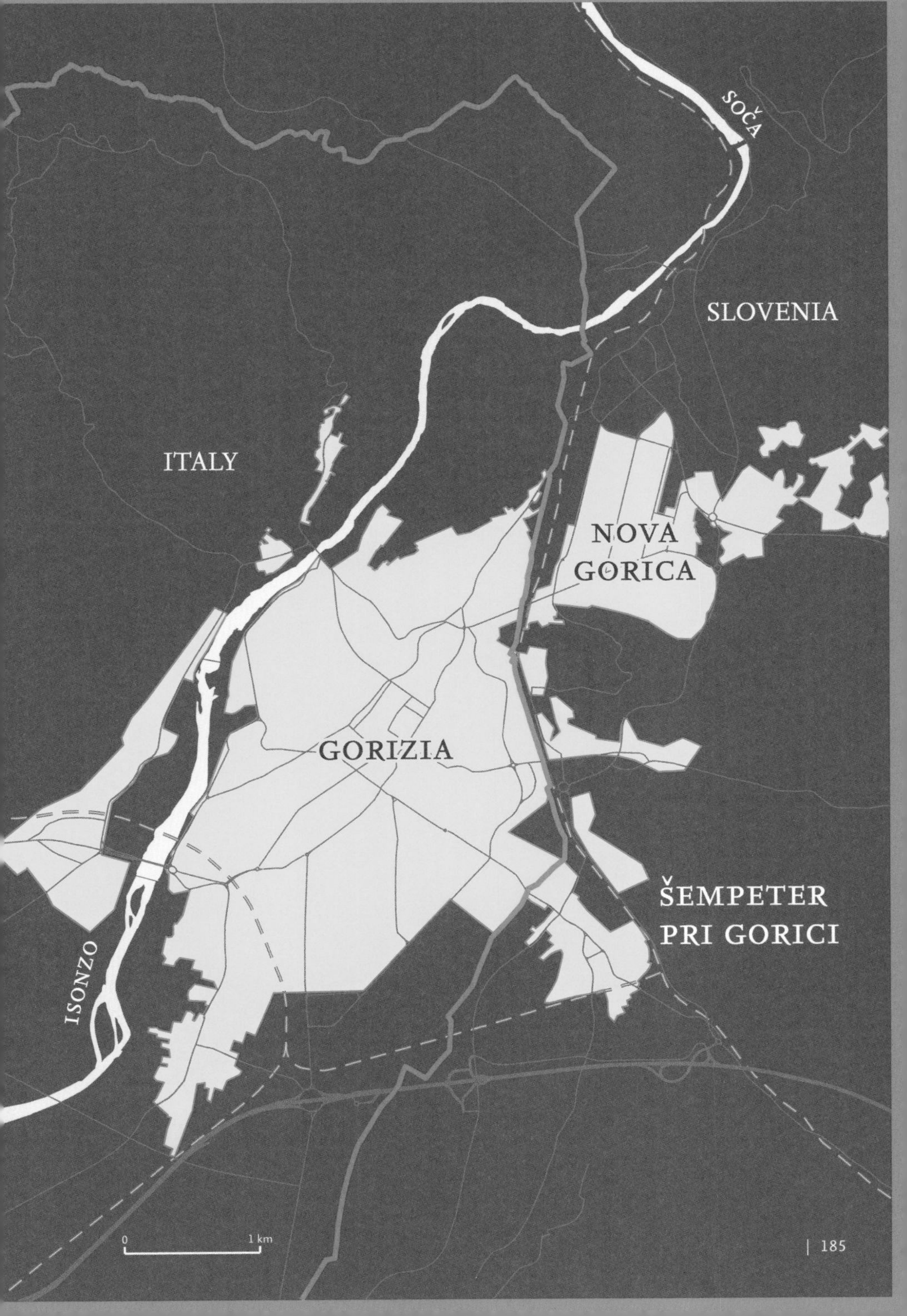

SOČA

SLOVENIA

ITALY

NOVA GORICA

GORIZIA

ŠEMPETER PRI GORICI

ISONZO

0 1 km

The valley of the Soča River (Isonzo in Italian), whose source is in the Alps and which flows down to the north of the Adriatic Sea, has always been a pleasant place to live, well known for the production of quality wines. A great part of this region is sheltered from a cold north wind – known as the bora – by hills and mountains. As the valley is open towards the Adriatic Sea to the south, the flow of warm air provides a mild Mediterranean climate. Therefore, it is no wonder that a village was built in this valley back in the tenth century; it was named **Gorizia** (in Slavic languages this means 'little hill'). During the Habsburg rule, from the sixteenth century, Gorizia developed rapidly and became an important multi-ethnic town in which numerous languages (Friulian, Venetian, German and Slovene) were spoken. Later, at the beginning of the nineteenth century, Gorizia became a popular tourist destination for the Austrian nobility, earning it the nickname of 'Austrian Nice'.

During the First World War, Italy fought on the side of the Allies, and fierce battles were held between the Italian and Austro-Hungarian armies around Gorizia. When the war was over, Gorizia became a contentious territory wanted by the short-lived State of Slovenes, Croats and Serbs (later part of the Kingdom of Yugoslavia) and, at the same time, by the local Friulians, who planned to remain as an autonomous region under Habsburg rule. This dispute was ended when the Italian army took control of Gorizia.

Between the two world wars, Gorizia remained within Italy, and extreme Italianization of Slovenians was carried out, with complete prohibition of the use of the Slovene language (even the names on the graves were Italianized).

During the Second World War, Gorizia was liberated by the Yugoslav Partisans, but was handed over to the Allies. After a few years, it was decided that Gorizia should belong to Italy, and its northeastern suburbs and surrounding villages should belong to Yugoslavia, namely Slovenia. The Yugoslav and Slovenian authorities immediately decided to build a new town, under the name Nova Gorica

('New Gorizia'), right alongside Gorizia and the border with Italy, which would become the centre for the newly acquired suburbs and villages.

Nova Gorica was quickly built in the following years on the principles of modernist architecture. Construction began in 1948 and as early as 1952 it was officially proclaimed a town, to which the surrounding settlements were joined.

Relations between Italy and Yugoslavia, and later Slovenia, regarding Gorizia were mainly good. This included numerous sports and cultural events, which strengthened the spirit of unity in the divided town.

In recent times, after Slovenia joined the European Union and the Schengen Area, the border dividing the town was completely lifted and free movement was allowed. The next step, the gradual merging of Gorizia and Nova Gorica, has already begun: a joint administration board has been established, which will manage the trans-border urban region. This region consists of Gorizia, Nova Gorica and the Slovenian town of Šempeter pri Gorici (San Pietro in Italian), a former suburb of Gorizia. These three towns already represent a continuous urban settlement, so their merging is a logical step forward.

Note: Friulian or **Friulan** is a Romance language, a sub-group of the Rhaeto-Romance languages. It is spoken in the Friuli region of Italy (Friuli–Venezia Giulia region), in the vicinity of Udine, Pordenone and Gorizia, near the borders with Slovenia and Austria. Friulian is spoken by about 600,000 people. The majority also speak Italian. The language is partially used in local schools and administration. The **Venetian language** is spoken by over two million people in Veneto, from Venice to Verona in the west, south to the River Po and east to Friuli and Trieste. Venetian is spoken in the littoral area of Slovenia and in Istria and somewhat less in Dalmatia in Croatia (about 50,000 people speak this language). This language is also spoken in Brazil (with more than a million speakers) and Mexico.

Border Mark, Gorizia | Nova Gorica, Italy | Slovenia

HERZOGENRATH AND KERKRADE

GERMANY | NETHERLANDS

The European town with a rebellious spirit

50º 51' 46"N | 06º 04' 47"E

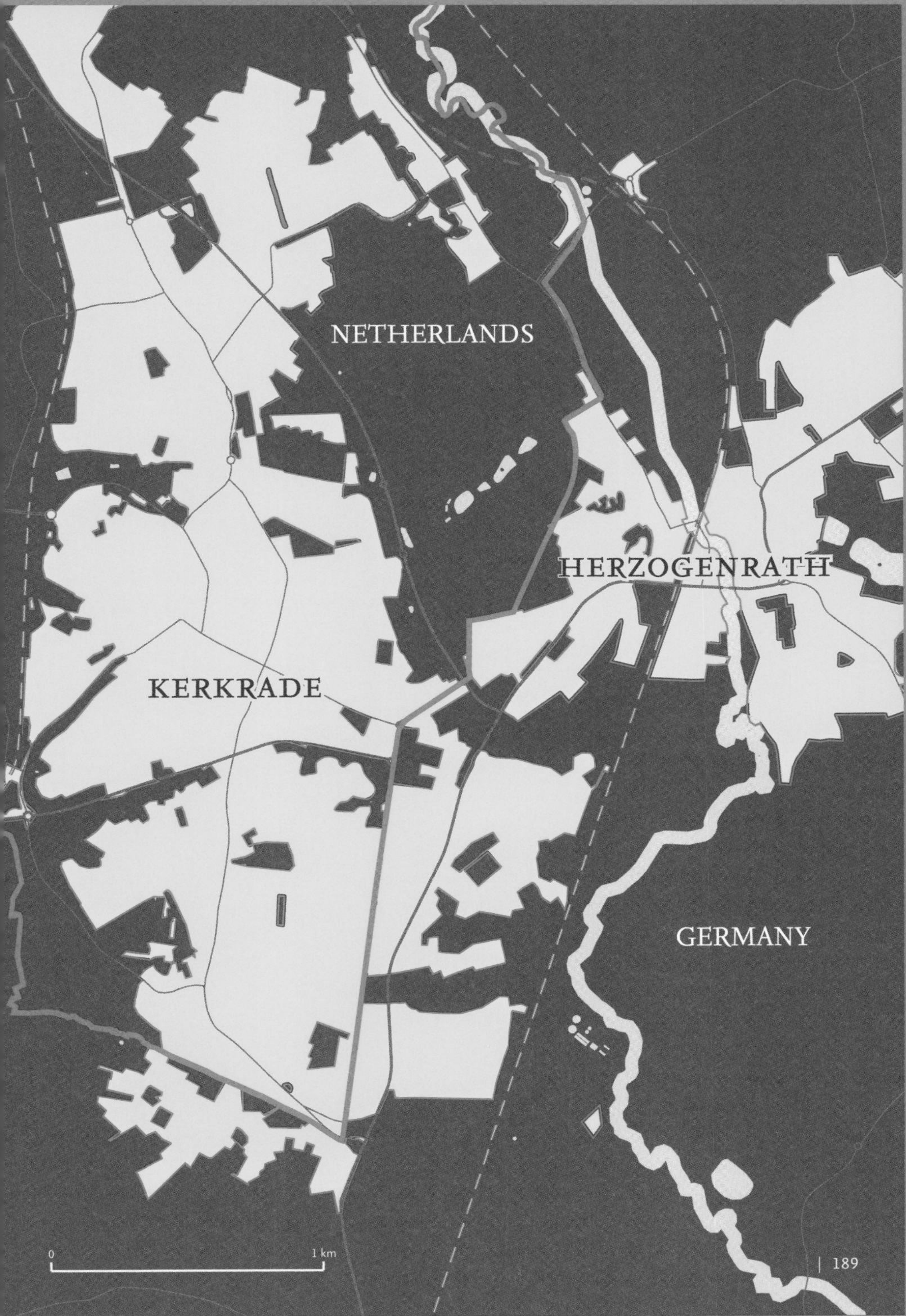

NETHERLANDS

HERZOGENRATH

KERKRADE

GERMANY

0

1 km

On one side of the main street is the town of Kerkrade, and the country of the Netherlands; a few metres away, on the other side of the street, is the town of Herzogenrath, and the country of Germany. The boundary between them is not visible, although it exists, at least officially.

Herzogenrath is a town in the district of Aachen in the German state of North Rhine-Westphalia. The town was founded in the eleventh century, under the name of Rode. During its 1,000 years of growth and development, Herzogenrath often changed ownership – the Spaniards were there in the seventeenth century, the Austrians in the eighteenth and the French up to 1813. The Kingdom of the Netherlands was formed in 1815 under the terms of the Vienna Congress. The border between the Netherlands and Prussia (Germany) was drawn through the centre of the town so the eastern part, called Herzogenrath, remained under German rule, while the western part, with the new name of Kerkrade, was granted to the Netherlands.

Kerkrade is in the far south of the Netherlands. At the beginning of the nineteenth century, it was an important mining centre, so it developed rapidly, absorbing all surrounding smaller towns. Today, the two towns each have approximately half of the total population of 100,000.

Life in the divided town was relatively simple. People on both sides of the new border spoke the same dialect, and the cultural unity was still felt. During the First World War, the border was clearly marked for the first time – with a two-metre-high fence. The fence's primary purpose was to prevent the German soldiers from deserting. Only after the Second World War did a gradual lowering of the fence begin: first, a considerably more pleasing wire fence was put in place, 120 centimetres in height. Later, in the 1960s, this fence was replaced with concrete pillars, sixty centimetres in height, without a fence between them. The next step was the removal of these pillars and the erection of a low wall down the middle of the road, twenty centimetres in height. The purpose of this was to prevent cars crossing from one side to the other. Pedestrians could normally step over the wall, which they often did – for quite some time, the smuggling of cheaper goods from one country to the other was quite common.

At the end of the twentieth century, the wall was removed completely. Until then there was a two-lane road on each side of the wall; the Dutch side of the street was called Nieuwstraat, and the German Neustrasse. The street is now a single two-way road with additional parking spaces, bicycle lanes and trees. The border is not marked, but drivers cross it whenever they overtake a vehicle. Public transport along the main street is under the competence of a German utility company from Aachen.

An unusual problem was once solved by rule-breaking at a local level. The German authorities had ordered the local authorities of the two towns to display both German and Dutch road signs, side-by-side on the traffic pillars. Though the German road signs had larger dimensions than the Dutch, they both displayed the same information. The authorities of Kerkrade and Herzogenrath considered this to be an unnecessary expense, and so decided to apply a more cost-effective solution – to only use the smaller (and cheaper) Dutch signs.

Today, the name Eurode is being applied more often for the community of Kerkrade–Herzogenrath, which is a term conceived from combining the word Europe with the historical region Land van 's-Hertogenrode.

Note: The only remains of a Roman villa in the Netherlands, available to tourists, are located in Kerkrade.

MARTELANGE

BELGIUM | LUXEMBOURG

Inaccurate maps caused the town to be divided

49° 50' 00"N | 05° 44' 30"E

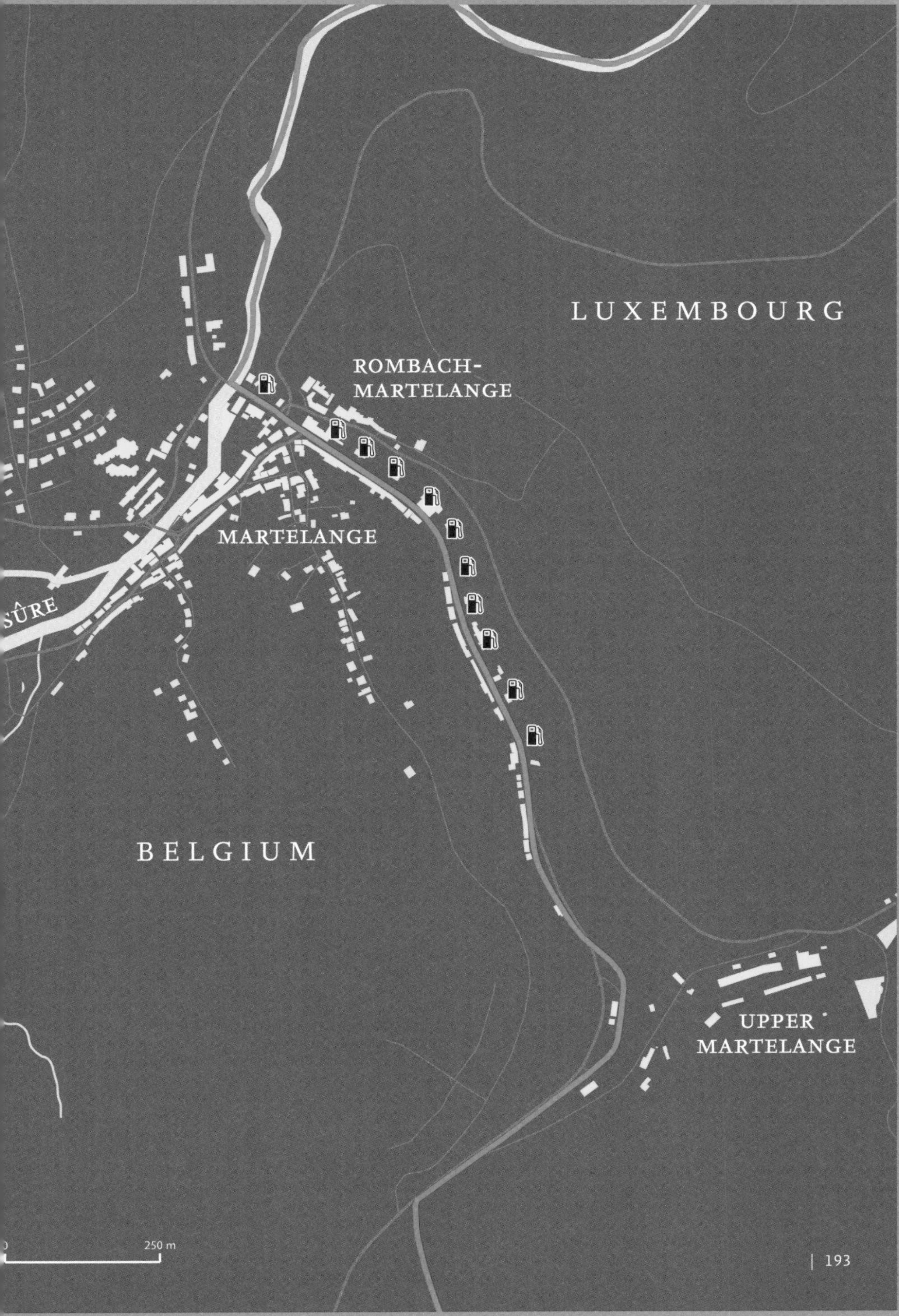

LUXEMBOURG

ROMBACH-
MARTELANGE

MARTELANGE

SÛRE

BELGIUM

UPPER
MARTELANGE

0 250 m

International borders that pass through towns very often cause some kind of problem for their inhabitants. However, the presence of a border may also benefit such a divided settlement, primarily by making it more interesting and therefore a tourist attraction, thus encouraging economic development in some way.

The town of **Martelange** is located in the southeast of Belgium, on the border with the neighbouring Grand Duchy of Luxembourg. A small part of this once-united town belongs to Luxembourg and is called Upper Martelange (Uewermaartelengin Luxembourgish; Haut-Martelange in French; Obermartelingen in German).

The town is primarily known for the N4 highway that runs through it, one of the most important connections between the capitals of the neighbouring countries (Brussels in Belgium and Luxembourg's Luxembourg). In fact, the N4 forms the border between Belgium and Luxembourg as it passes through the town. Since fuel, tobacco and alcohol taxes are considerably lower in Luxembourg than in Belgium, the Luxembourg side of the N4 is densely lined with petrol stations and small off-licences and tobacco stores, and not much else.

This unusual border situation dates back to the mid-nineteenth century, when the borders of the Kingdom of Belgium were determined. The border commission decided that the entire road, which is now known as the N4, should pass through Belgian territory in order to avoid border crossings. Unfortunately, old and inaccurate maps were used, which resulted in some parts of Martelange remaining within Luxembourg. The mayor at the time protested about the division, and although the protest was accepted, the Dutch (who were then in a union with the Grand Duchy of Luxembourg) had already put boundary markings in place, and so the division of Martelange was officially implemented.

Modern times have once again led to a kind of merging of the two sides of Martelange. Immediately after the Second World War, the union of Belgium, the Netherlands and Luxembourg (Benelux) was formed, and a little later the forerunners of the European Union evolved (the European Coal and Steel Community, the European Economic Community, the European Community). Each of these unions gradually decreased the significance of the borders, but the tax rates remained within the control of the individual member states. Today, the inhabitants of Martelange mainly get their fuel from one of the numerous petrol stations on the Luxembourg side of the N4. This side of the highway is now part of a small town, Rombach, but due to its connection with its Belgian 'twin', it is known as Rombach-Martelange.

Such a density of petrol stations (around fifteen in one kilometre) indirectly caused a major accident at the end of the 1960s. A tanker containing about forty-five tonnes of gas overturned, causing an explosion that killed twenty-two people and injured 120. More than twenty buildings, including a pharmacy, hotel, post office, store and bank, were completely destroyed. A memorial was erected in memory of the tragedy. Unfortunately, it too, suffered a misfortune: in 1990, another tanker fell into the river, causing significant chemical pollution and the collapse of the monument.

It is interesting to note that the border between Belgium and Luxembourg does not actually run down the middle of the N4 highway, as expected, but about a metre or two from the Luxembourg edge of the road. This means that almost all things on the Luxembourg side are, in fact, partly in Belgium. Martelange is also known as a four-language town: Walloon (the northernmost Romance language), French and Luxembourgish are spoken on the Belgian side; while Walloon, German and Luxembourgish are used on the Luxembourg side.

Martelange, Belgium | Luxembourg

ISTANBUL

TURKEY

The mega-city that spans a continental divide and almost 3,000 years of history

41° 01' 52"N | 28° 59' 59"E

BLACK SEA

BOSPORUS

ISTANBUL

BLUE MOSQUE

HAGIA SOPHIA

SEA OF MARMARA

0

10 km

The largest city in Europe, the largest city in the Middle East, and the sixth largest city in the world. A city almost 3,000 years old, the capital of several former world empires, and the world's fifth most popular city tourist destination.

The city that meets all these criteria is too large for a country, and even for a continent. That is why **Istanbul**, with a population of fifteen million and an area of about 5,000 square kilometres, spreads into two continents, Europe and Asia. This megalopolis is the cultural, economic and historical centre of Turkey. The commercial and historical centres are located in the European part of the city, while a third of the population lives in the Asian part. This transcontinental city has access to two seas (the Sea of Marmara and the Black Sea) and the strait that connects them (the Bosporus).

The first settlement in what is today's European part of Istanbul was founded in the seventh century BC, under the name Byzantium (Βυζάντιον in Greek), by Greek colonists from the Athens area. The significance of Byzantium rose rapidly in the fourth century, when Emperor Constantine the Great made it the new capital of the Roman Empire under the name of Constantinople (Κωνσταντινούπολι, 'City of Constantine'). Although the emperor attempted to promote the name New Rome, this did not enter widespread usage. Unofficially, Istanbul was usually simply called the City. After the Ottoman conquest of Constantinople in 1453, up until the proclamation of the Turkish Republic in 1923, the name Kostantiniyye was used, meaning 'Constantine's city'. The name Istanbul was first mentioned in the tenth century in Armenian and Arabic sources. This name, presumably, comes from the Greek phrase 'εἰς τὴν Πόλιν' (pronounced eis tēn pólin), which means 'to the city'. It became the official name only after the proclamation of the republic in 1923.

Today, Istanbul is characterized by its high-density population, as well as numerous monumental buildings, such as the well-known Hagia Sophia, built during the reign of the Eastern Roman emperor (Byzantine), Justinian I, in the sixth century. The Blue Mosque (Sultan Ahmed Mosque) stands out for its grandiosity, as well as the magnificent Dolmabahçe Palace with its 285 rooms and forty-six richly decorated halls.

The Asian part of Istanbul consists of several commercial zones, large residential areas, big parks and marinas, as well as the only warm water spring within the territory of Istanbul. The Asian part is connected to the European part by two bridges and a railway tunnel. Work on the tunnel was delayed numerous times and deadlines extended because workers kept coming across valuable historical findings. The construction of a third bridge and second tunnel is planned.

THE URAL RIVER AND CITIES ON TWO CONTINENTS

The river that splits towns between two continents along its 2,500-kilometre course

46° 59' 35"N | 51° 44' 49"E

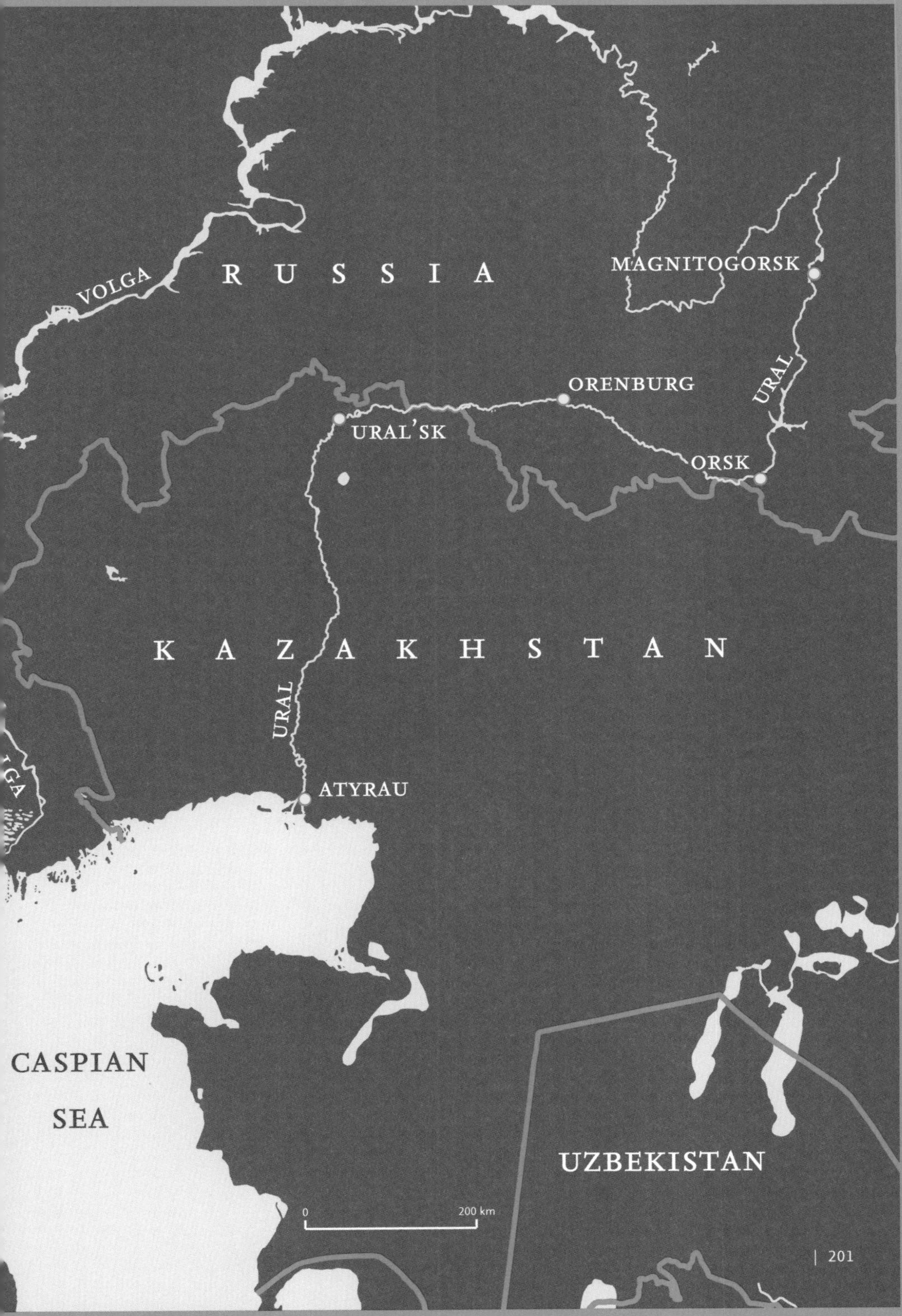

VOLGA
R U S S I A
MAGNITOGORSK
URAL
ORENBURG
URAL'SK
ORSK
K A Z A K H S T A N
URAL
ATYRAU
CASPIAN
SEA
UZBEKISTAN
0
200 km

The Ural River rises in the southern slopes of the Ural Mountains and flows into the Caspian Sea. Its total length is almost 2,500 kilometres, which makes it the third-longest river in Europe and the eighteenth-longest river in Asia. The river's source is in the Russian Republic of Bashkortostan (formerly Bashkiria), where this powerful water flow is known as the Yaik, and its mouth is in Kazakhstan, whose local population call it the Jaiyq. The Russian name of this river was Yaik until the end of the eighteenth century.

Most experts consider this river to form part of the Europe-Asia boundary. Viewing the whole boundary, it passes through the Ural Mountains, the Ural River, the Caspian Sea, the Caucasus Mountains, the Black Sea, the Bosporous, the Sea of Marmara and the Dardanelles. Over time, many cities that had formed on one bank of the Ural River have gradually extended to the other bank, and, therefore, to another continent.

The first city of that kind is **Magnitogorsk**, an industrial city located in the extreme south of the Ural Mountains, on the banks of the river. Although its population has been decreasing in recent years, it still has over 400,000 inhabitants. Magnitogorsk Iron and Steel Works (MMK) is the main industrial plant in this city, being one of the largest steel works in Russia. Proximity to large iron ore deposits provided the reason for the town's development, which was of crucial importance during the Second World War. Magnitogorsk was a long way from the front line, which made it possible for the city to produce large quantities of much-needed steel. Of course, such mass production caused severe pollution of the entire area, and the problem still exists – Magnitogorsk is among twenty-five of the world's most polluted cities. Depletion of the available quantity of iron ore forced MMK to import it from Kazakhstan.

Some 250 kilometres south of Magnitogorsk lies the city of **Orsk**, another two-continent city on the Ural River. Today Orsk has about 240,000 inhabitants. It was a significant trade centre during the eighteenth century, but was also known for the manufacture of top-quality shawls and scarves, whereas in the twentieth century it was well known for the mining of a remarkable semi-precious stone – jasper.

Orenburg, the capital of Orenburg region, 250 kilometres further downstream from Orsk, is another city that extends onto two continents, with its bigger part being in Europe, and the smaller part in Asia. Orenburg is located less than 100 kilometres from the border with Kazakhstan. Its population is over half a million, and it is a large centre for the Russian energy sector. In addition, this city has a number of universities, institutes, museums and theatres, which make Orenburg a significant regional educational and cultural centre as well.

About eighty-five kilometres (in a straight line) from the point where the Ural crosses the Russia–Kazakhstan border, the river passes through the city of **Ural'sk**, the first transcontinental city of Kazakhstan. Most of this city is located to the west of the Ural, making it a mostly European city. Some small settlements and the airport are situated on the opposite bank, in Asia. This city, as well as the majority of other cities on the Ural River, was founded by Cossacks in the seventeenth century. Nowadays it is an important agricultural and industrial centre, and provides a connection between the oil fields in the Caspian region and the industrial cities of the southern Ural.

At the very mouth of the Ural River in the Caspian Sea, is the city of **Atyrau**, known as Gur'yev until 1991. This city with about 155,000 inhabitants is now known for its oil industry and fishing. In addition, it is one of the two main Kazakhstan ports on the Caspian Sea, and is located about twenty metres below sea level. Some of the most significant Kazakhstan oil pipelines to Russia start from this city.

GHOST TOWNS

PRYP"YAT'
PLYMOUTH
KOLMANSKOP
CENTRALIA
CRACO
HASHIMA ISLAND

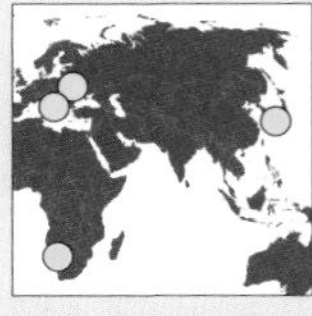

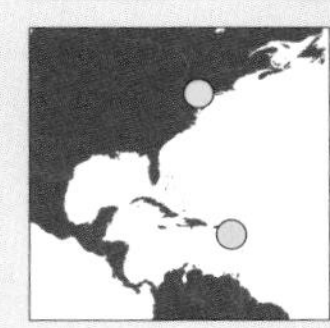

Natural disasters, nuclear emergencies, economic decline and a fire in a coal mine have all caused residents to flee

Pryp"yat' 51° 24' 20"N | 30° 03' 06"E
Plymouth 16° 42' 22"N | 62° 12' 40"W
Kolmanskop 26° 42' 13"S | 15° 13' 54"E
Centralia 40° 48' 13"N | 76° 20' 28"W
Craco 40° 22' 37"N | 16° 26' 28"E
Hashima Island 32° 37' 39"N | 129° 44' 17"E

Pryp’yat’, Ukraine

The term 'ghost town' means an abandoned village, town or city, usually with substantial visible remains. The reasons these settlements were abandoned can be numerous, the most common one being a decline in economic activity. Other reasons may include natural and other disasters (floods, wars, uncontrolled anarchy, and incidents in large industrial facilities).

Sometimes the term 'ghost town' is also used for settlements that have not been completely abandoned, but whose population has significantly decreased. Many such towns become tourist attractions, especially if a large number of buildings are still intact.

One well-known ghost town is **Pryp"yat'**, a town in northern Ukraine, near the border with Belarus. The town was founded in 1970 for the needs of the employees of the nuclear power plant 'V. I. Lenin' near Chernobyl. At the time of the 1986 nuclear explosion in Chernobyl, Pryp"yat' had about 50,000 inhabitants. They were all urgently evacuated after the disaster, leaving whatever it was they were doing (children's toys were left in parks, laundry was left drying on clothes lines, food was left on stoves and in fridges...). Since then, most of the town has been robbed, and everything that could be taken away, has been (furniture, tools, even the toilet bowls). People are not allowed to live within a radius of thirty kilometres from the plant, but visitors are allowed to see the town, as a kind of an open-air museum. Plants and animals are specially monitored for the impact of radioactivity upon them.

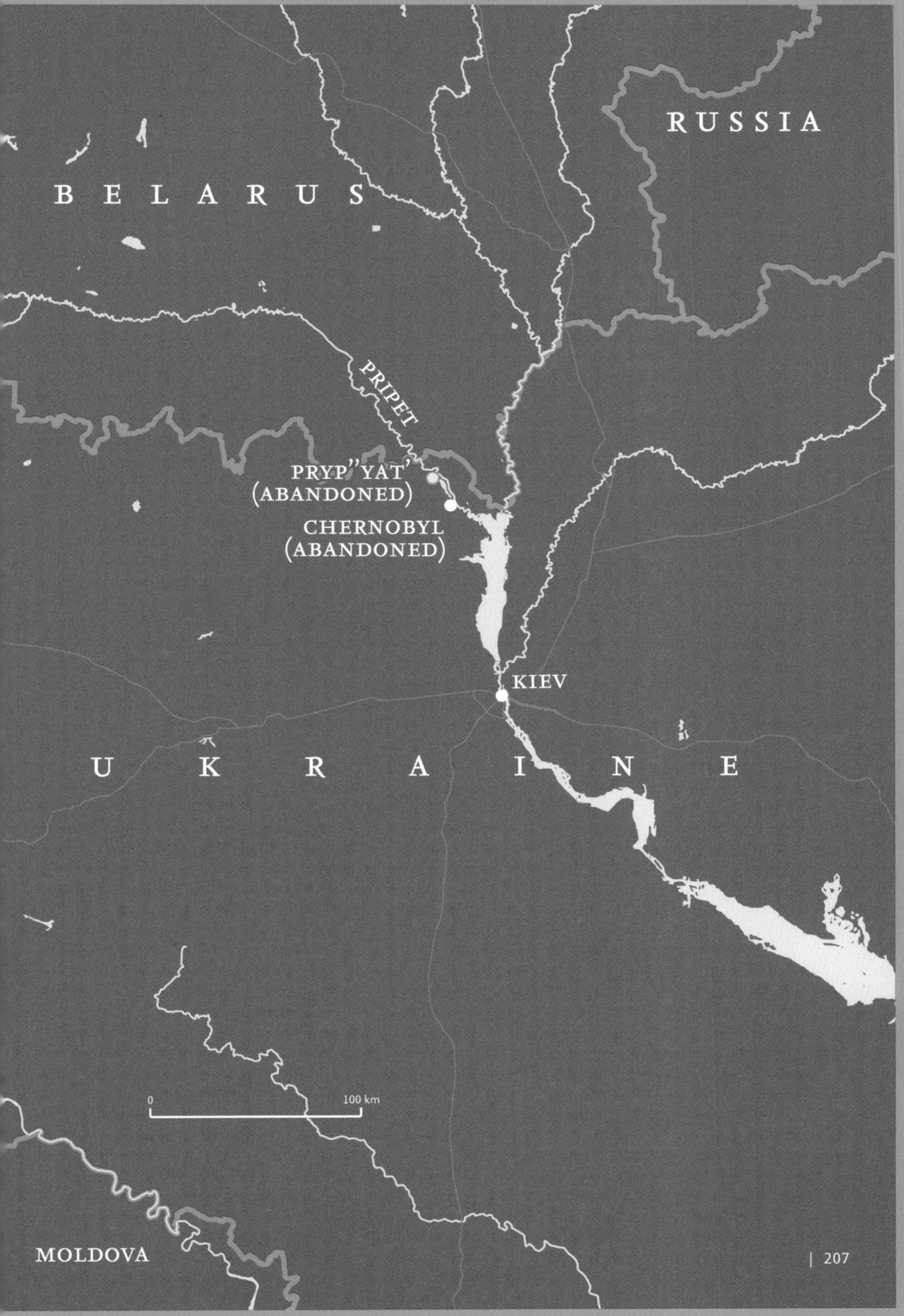

RUSSIA
BELARUS
PRIPET
PRYP"YAT'
(ABANDONED)
CHERNOBYL
(ABANDONED)
KIEV
UKRAINE
0
100 km
MOLDOVA

Officially, **Plymouth** is still the capital of the island of Montserrat, the British Overseas Territory in the Caribbean Sea, although it was completely evacuated after the volcanic eruption of 1995, which practically destroyed the southern half of the island. The local government was evacuated to the northern settlement of Brades, while the construction of a new capital started in the vicinity, at the location known as Little Bay. The eruption caused huge damage – Plymouth, the biggest town and port, was covered in ash and mud, the airport was destroyed, and two-thirds of the population was evacuated to other islands, where most of them have remained.

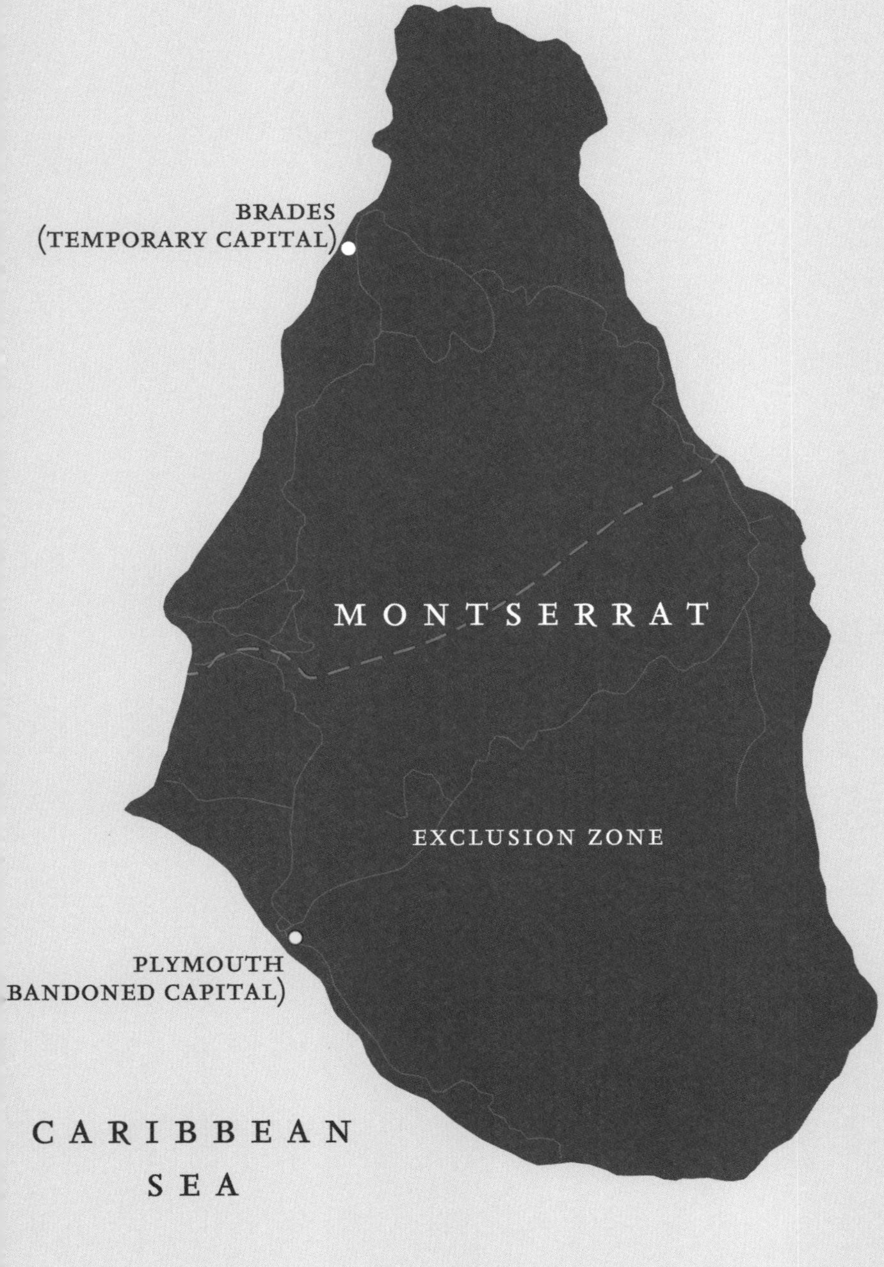

BRADES
(TEMPORARY CAPITAL)

MONTSERRAT

EXCLUSION ZONE

PLYMOUTH
BANDONED CAPITAL)

CARIBBEAN
SEA

0
2 km

Another ghost town, in the Namib Desert in southern Namibia, is **Kolmanskop**. This little town was founded in the early twentieth century, when diamonds were discovered nearby. The importance of Coleman's Head (a translation of its name from Afrikaans, the language of the Dutch conquerors in southern Africa) decreased along with the quantity of available diamonds, and the town became completely abandoned in the middle of the twentieth century. This ghost town, with its German-style houses that are slowly being buried by sand, is now a popular tourist attraction, only some fifteen kilometres away from the Namibian port of Lüderitz.

Kolmanskop, Namibia

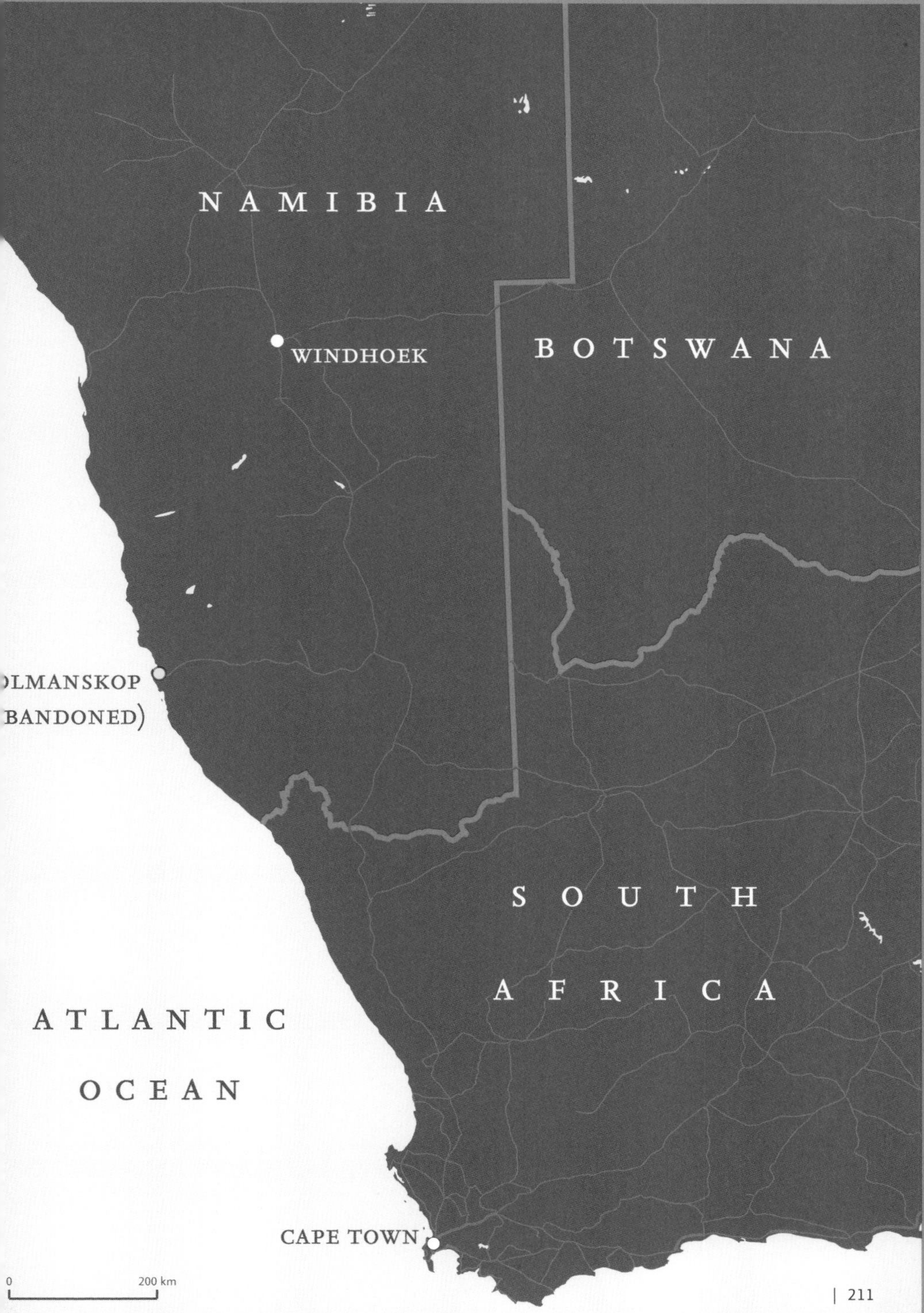

NAMIBIA

WINDHOEK

BOTSWANA

LMANSKOP
BANDONED)

SOUTH
AFRICA

ATLANTIC
OCEAN

CAPE TOWN

0
200 km

The little town of **Centralia** in Pennsylvania is almost a ghost town. Its population dropped from 2,500 after the Second World War to about 1,000 in the early 1980s, to a dozen people in 2013. The reason for such a rapid drop in the number of inhabitants is the underground fire that has been continuously burning since 1962 in the Centralia coal mine, and which experts predict will burn for another 250 years. As a consequence of the fire, toxic gases are rising from the ground all over the town, the most dangerous being carbon monoxide (a gas without colour, taste or smell, which can be deadly when inhaled). Another hazard is the sudden opening of large sinkholes and the collapsing ground affecting buildings and roads. The government of Pennsylvania made a decision to urgently buy the entire town from its owners, and to immediately have most of the buildings pulled down. Only a dozen very persistent inhabitants have remained in their homes. The nearby, considerably smaller, town of Byrnesville has shared Centralia's fate.

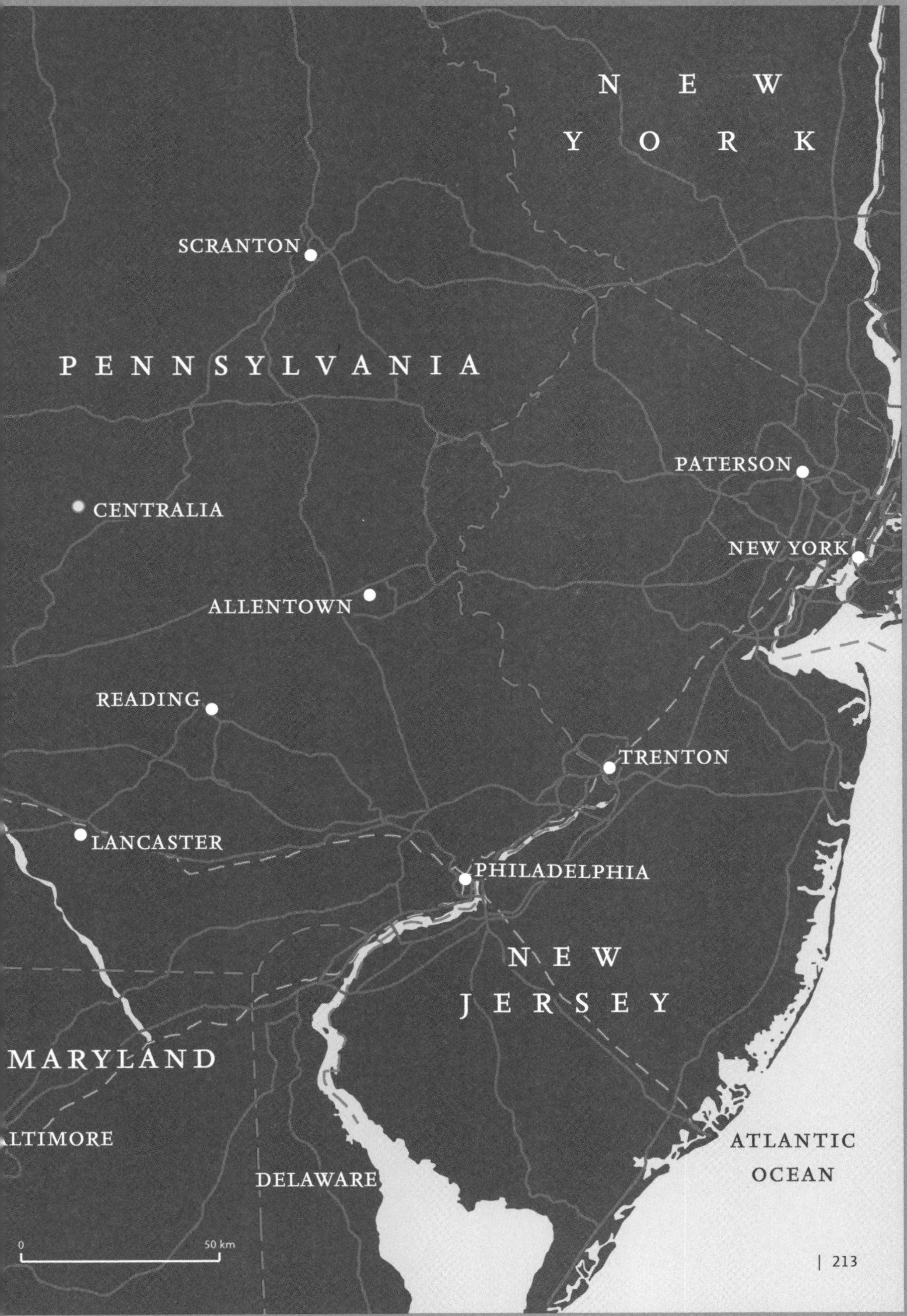

NEW YORK
SCRANTON
PENNSYLVANIA
PATERSON
CENTRALIA
NEW YORK
ALLENTOWN
READING
TRENTON
LANCASTER
PHILADELPHIA
NEW JERSEY
MARYLAND
LTIMORE
ATLANTIC OCEAN
DELAWARE
0
50 km

One of the most famous ghost towns is **Craco**, a popular filming location due to its striking appearance. Craco is located in the south of Italy, and for defensive reasons was built on a 400-metre-high, steep summit, above the valley of the Cavone River. The first evacuations of the population started in the 1960s, due to the landslide that threatened the city. It seems to have been provoked by human activity, in particular the construction of a sewerage and water supply system. A flood in 1972 further aggravated the situation, and an earthquake in 1980 led to the final abandonment of the town.

Craco, Italy

CAMPOBASSO
ADRIATIC SEA
BARI
ITALY
NAPLES
POTENZA
GULF OF SALERNO
CRACO
GULF OF TARANTO
TYRRHENIAN SEA
0 50 km
CATANZARO
IONIAN SEA
SICILY

The Japanese **Hashima Island** is one of the most attractive ghost towns, also known as Gunkanjima (Battleship Island). This island was inhabited from 1887 to 1974, when it was used as a coal mine. The decline in demand for coal resulted in the closure of the mine, and in 2002 its owner, the Mitsubishi Company, donated the island to the city of Takashima, which is currently part of Nagasaki. The island is interesting because of the large number of buildings made of concrete (as early as 1916, a nine-storey building was constructed from concrete due to typhoon hazards, and served as accommodation for the mine workers). Hashima is popular among tourists and admirers of old buildings, though only one part of the island is open to visitors – a lot of money would be needed to secure ruined buildings and make the rest of the island safe for vistors.

ULSAN
SOUTH KOREA
BUSAN (PUSAN)
JAPAN
HIROSHIMA
YAMAGUCHI
TSUSHIMA
IKI-SHIMA
FUKUOKA
SAGA
ŌITA
GOTŌ-RETTŌ
KUMAMOTO
NAGASAKI
HASHIMA ISLAND
KYŪSHŪ
MIYAZAKI
KAGOSHIMA
PACIFIC OCEAN
AST CHINA SEA (DONG HAI)
TANEGA-SHIMA
YAKU-SHIMA
0
100 km

JEWISH AUTONOMOUS OBLAST

RUSSIA

The success story of the people who moved from one side of Russia to the other

48° 39' 23"N | 132° 17' 15"E

RUSSIA

KHABAROVSKIY KRAY

MURSKAYA
OBLAST'

BIROBIDZHAN

AMUR

KHABAROVSK

JEWISH AUTONOMOUS OBLAST

AMUR

HEILONG JIANG

HEILONGJIANG

HEGANG

CHINA

JIAMUSI

SHUANGYASHAN

PRIMORSKIY

KRAY

QITAIHE

JIXI

LAKE
KHANKA

0 100 km

If you looked at a map of the Far East, north of Japan, you would notice the large Russian island of Sakhalin. If you focused in on the northwestern part of this island, you would notice that opposite the island, on the coast, is the mouth of the powerful Amur River. If you continued to 'sail' upstream on the Amur, after several hundred kilometres you would come across the large Russian city of Khabarovsk, as well as two borders near to that city – the Russia–China border, and the border of Russia's only autonomous oblast (province), the **Jewish Autonomous Oblast** (JAO).

What is a Jewish Autonomous Oblast doing in the Far East? It is well known that the ancient Romans displaced a large number of Jews from Palestine; the people found a new home in Europe, primarily in the south (Spain), but also in Central and Eastern Europe. Many of those Central European and East European Jews eventually found themselves within the boundaries of the Russian Empire, later the Soviet Union.

In the mid-1930s, the Soviet authorities, headed by Stalin, made a decision to give many racial and ethnic groups within the territory of the USSR the opportunity to form autonomous communities, where they would be able to freely develop their culture, language and tradition within Soviet frameworks. Some of these nations got their autonomous provinces in the same locations where the majority of their people were already living (for example, the German Soviet Republic was formed in the lower course of the Volga River, where the majority of Russian Germans lived). However, others were relocated, either willingly or by force, to territories selected by the supreme government. The latter case applied to the Jews who, at the time of establishment of JAO, had mostly been settled in the European part of Russia (presently Ukraine, Belarus, the Baltic countries and western Russia).

A remote wetland along the Amur River bank, opposite China's Manchuria, was selected to be the location for their autonomous province. This is where the foundations of the 'Soviet Zion' were laid, where the development of the Jewish proletarian culture was envisioned, as a sort of gateway to religious and nationalist Zionism. In order to develop this plan more easily, the use of Yiddish was enforced in JAO (essentially the German language greatly influenced by Slavic languages, which was used by the Jews in Central Europe), instead of the religiously influenced Hebrew language. When this province was selected as the location of JAO, it was almost completely uninhabited. All towns, including the capital Birobidzhan, were therefore founded in the mid-1930s, although the first Jewish settlers had begun to arrive in this area some ten years earlier.

Despite being strongly promoted, it was a relatively small number of Jews that decided to move across the country to the Oblast. This meant that there were always relatively few Jewish settlers there – the largest population was reached after the Second World War, but even then only a quarter of the total number of inhabitants were of Jewish origin. Though not numerous, the Jewish community was strongly present in JAO at a cultural level: they had many Jewish schools, theatres, newspapers and significant writers.

Today, JAO is one of the wealthiest regions of Russia, with highly developed industry and agriculture, as well as a dense traffic network. The status of a free trade zone, considerable mineral wealth, significant areas of high quality forestation and food production all contribute to its success, too.

The people who gave their name to this autonomous region now make up less than 2 per cent of the population. However, Yiddish is still being taught in several villages, where Jewish people are in the majority, and the most prominent local newspapers still have several pages printed in this language.

BRITISH CROWN DEPENDENCIES

ISLE OF MAN

JERSEY AND GUERNSEY

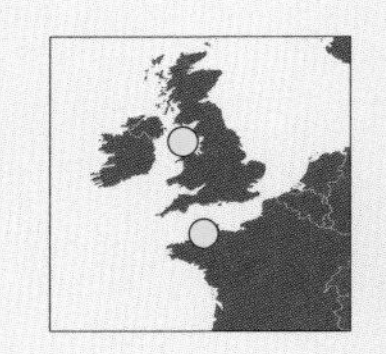

The island with the oldest continuous parliament in the world

Isle of Man 54° 15' 14"N 4° 31' 32"W
Channel Islands 49° 29' 26"N 2° 11' 53"W

SCOTLAND

SOLWAY FIRTH

ISLE OF MAN
(BRITISH CROWN DEPENDENCY)

ENGLAND

NORTHERN
IRELAND

DOUGLAS

I R I S H   S E A

ANGLESEY
(YNYS MÔN)

WALES

0 20 km

The British Isles archipelago contains two sovereign states: the United Kingdom of Great Britain and Northern Ireland (UK), and the Republic of Ireland. The first of these consists of four entities – England, Scotland, Wales and Northern Ireland; each of these entities has a certain level of self-government but, along with many of the islands that surround them, together they form the UK. There are many small islands around the UK. However, some of these islands differ from all the others, as they are not part of the United Kingdom, or the European Union. They are known under the common name of British Crown dependencies.

Crown dependencies are island territories, self-governing possessions of the British Crown, and as they do not form a part of the United Kingdom, they differ considerably from British Overseas Territories. Each of these islands differs in terms of the degree to which they are connected to the UK, and almost all laws and regulations, other than those concerning defence and international representation, are made by the local governments. The British government does not have any influence on the Crown dependencies, except if approved by them (in fact it has recently made a decision to intervene even less in their foreign affairs), while customs and immigration services are completely in the hands of the Crown dependencies.

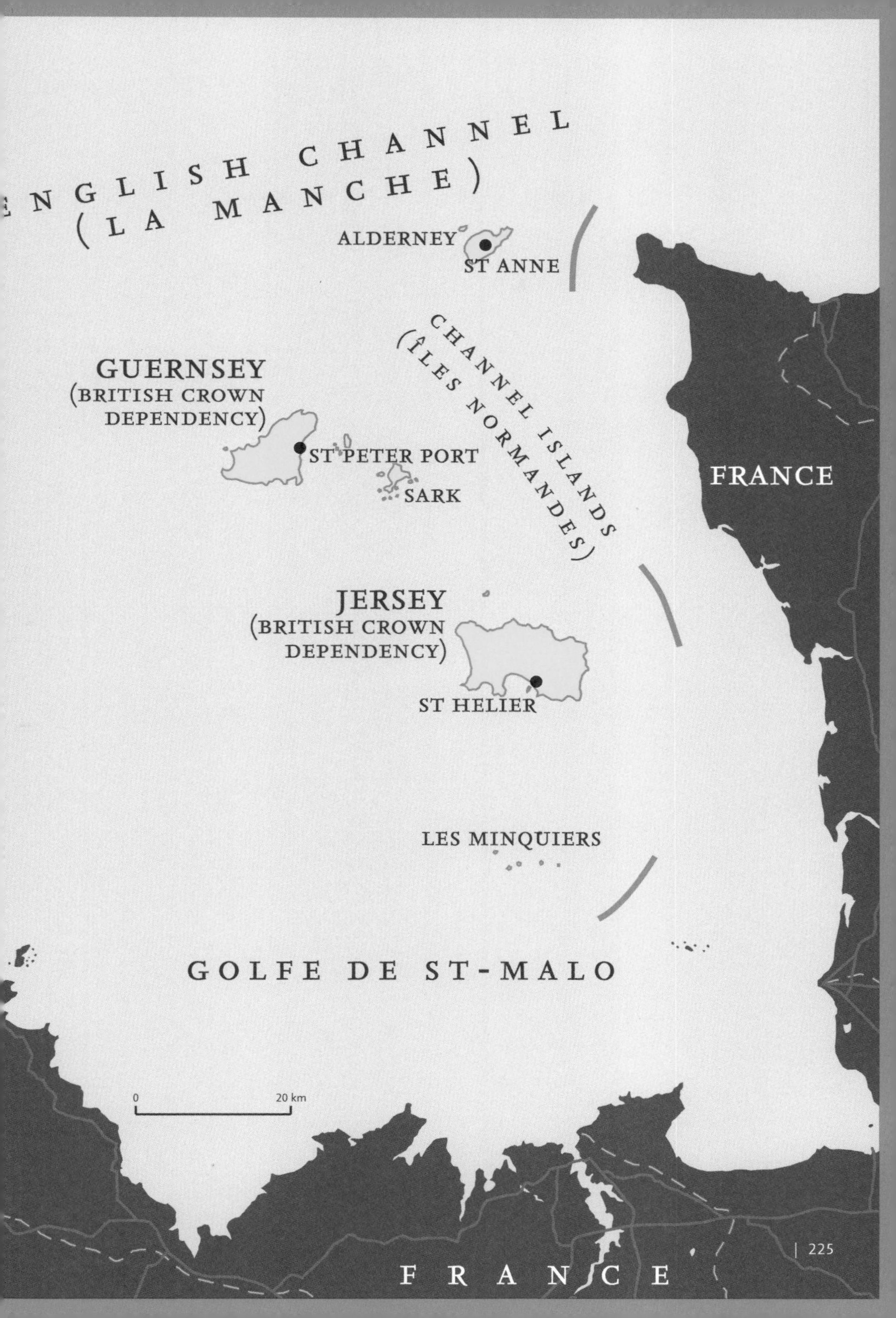
ENGLISH CHANNEL
(LA MANCHE)
ALDERNEY
ST ANNE
CHANNEL ISLANDS
(ÎLES NORMANDES)
GUERNSEY
(BRITISH CROWN
DEPENDENCY)
ST PETER PORT
SARK
FRANCE
JERSEY
(BRITISH CROWN
DEPENDENCY)
ST HELIER
LES MINQUIERS
GOLFE DE ST-MALO
0
20 km
FRANCE

Currently, the following islands have the status of Crown dependency:

The **Isle of Man.** Located in the Irish Sea, between Great Britain and Ireland, it has an area of over 570 square kilometres and a population of about 80,000. It is not part of the European Union, but it has an agreement on the free flow of goods. A customs union exists between the UK and the island, and due to the very low taxes there, many companies register their headquarters on the Isle of Man. The head of state is the British monarch, who holds the title Lord of Mann (the title is always the same, regardless of whether the British monarch is a king or queen). Tynwald, the island's parliament, is thought to be one of the oldest in the world according to many historians (founded at the end of the tenth century, only forty or so years after the oldest, Icelandic, parliament) and the first national legislative body in the world to give women the right to vote in a general election. Approximately half of the population is of Celtic origin. The local Gaelic language is Manx, which is greatly influenced by English and Nordic languages; though it is spoken by less than 2,000 people, much is being done to revive it.

Jersey. Located in the English Channel, the island of Jersey is the largest of the Channel Islands. The name Channel Islands relates only to a geographical entity – it is not a political one. The archipelago is about twenty kilometres from the coast of Normandy, France. It has several islands, some of which are uninhabited.

Jersey is not part of the UK, but the UK is constitutionally responsible for the defence of Jersey, and its special relationship with the EU allows free trade even without full membership. The British monarch reigns in Jersey with the title Duke of Normandy, and is represented on the island by a governor. There are about 100,000 people living on this island of about 120 square kilometres. They mainly speak English, though a smaller number speaks Jèrriais, a form of Norman language from nearby Normandy.

Guernsey. The smaller of the two Crown dependencies in the Channel Islands, Guernsey has an area of about seventy-eight square kilometres and a population of 65,000, out of which only about 2 per cent fluently speaks Guernésiais, which, just like Jèrriais, is a variation of Norman. The Guernsey Crown dependency, apart from the island of the same name, also includes the inhabited islands of Sark and Alderney – each with its own parliament – as well as a few surrounding uninhabited islands. Sark and Alderney, although part of Guernsey, have a wide autonomy, both in relation to Guernsey and the UK. The legal system is quite complicated and intertwined, due to the legacy of its feudal past, although many consider Sark to be the last feudal country or territory in Europe. Jersey and Guernsey, as part of the Channel Islands, represent the last remnants of the medieval Duchy of Normandy and the only territory of the UK that the Germans occupied during the Second World War.

Fort Grey, Guernsey

GERMAN GREEN BELT

The political and ideological divide that provided a physical haven for nature

51° 40' 05"N | 11° 27' 37"E

DENMARK
NORTH SEA
BALTIC SEA
HAMBURG
POLAND
MSTERDAM
THERLANDS
BERLIN
HANOVER
GERMANY
LEIPZIG
GIUM
FRANKFURT
PRAGUE
LUXEMBOURG
LUXEMBOURG
CZECHIA
(CZECH REPUBLIC)
NUREMBERG
FRANCE
LIECHTENSTEIN
VADUZ
AUSTRIA
BERN
SWITZERLAND
ITALY
0
100 km

For almost forty years the 'Iron Curtain' divided Europe into the West and the East. That unofficial, yet exceptionally strong, border stretched from the Norwegian–Russian border in the far north of Europe to the Bulgarian–Turkish border on the Black Sea. It was an insurmountable political, ideological and physical barrier, which was especially pronounced in divided Germany. The boundary between East and West Germany consisted of metal fences, walls, barbed wire, watchtowers and minefields. In order to monitor this border more effectively, the East German authorities requested a relatively wide zone around the borderline itself, which became a sort of 'no man's land'. As a consequence, there was a corridor from the Baltic Sea to the tri-border area of East Germany, West Germany and what was then Czechoslovakia (now Czechia and Slovakia), about 1,400 kilometres long and a few hundred metres wide, in which there was practically no human activity – no cultivation, no woodland management, and, with the exception of the border guards, no people passing through this territory.

Nature exploited this unusual situation and soon took over the entire 'Iron Curtain' zone between the two German states. After the reunification of Germany, the border was removed, and with the help of numerous ecological organizations, the German federal government and German provincial authorities, the Iron Curtain was replaced with the **Green Belt**. The intention was to use as much of the former border as possible as a protected natural reserve, with some possible additions of nature parks close to the Belt.

This goal has been fulfilled to a large extent. Some parts of the Belt had to be returned to the original owners; there were parts that were used for new roads (Germany has one of the world's densest road networks); and large industrial facilities along other parts of the Belt made it impossible to use these parts as natural ground. Nevertheless, more than half of the former Iron Curtain has been successfully included in the Green Belt, in addition to which significant areas have been added, donated by nearby German provinces, the federal government and various non-governmental organizations.

Today, the Green Belt is the habitat for numerous species of birds, animals and plants, and has become a significant tourist destination for lovers of untouched nature. An additional bonus is the fact that it enables the natural features of the three main types of German landscape to be seen: the coasts in the north, the plains in the central part and the low mountains in the south.

The German Green Belt is now a part of the future **European Green Belt** initiative, which will extend down the entire route of the former Iron Curtain corridor, with the slogan 'Borders separate, nature connects'.

Apart from Germany's Green Belt, there are numerous similar cases around the world, where so-called involuntary parks have been created. The reasons for the existence of such unofficial nature reserves are varied. Some are separation zones around or between certain border areas, such as: the **Cyprus Green Line** under the control of the United Nations; the demilitarized zone on the border between **North and South Korea**; and the border zone between **Hong Kong** and mainland **China**. In addition there are large military ranges, including: the US army's **White Sands Missile Range** in southern New Mexico, which occupies almost 8,300 square kilometres; and **Monte Bello Islands**, an Australian archipelago with over 150 islands on which the British carried out nuclear tests in the 1950s. While other areas that are now left to nature are those that were hit by major disasters – earthquakes, tsunamis, industrial accidents – and are no longer suitable for humans.

German Green Belt, Germany

THE PRINCIPALITY OF ANDORRA

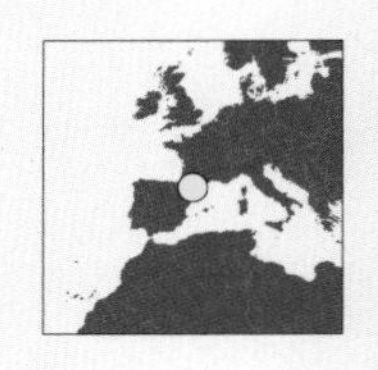

The independent Catalan nation governed by a bishop from Spain and the president of France

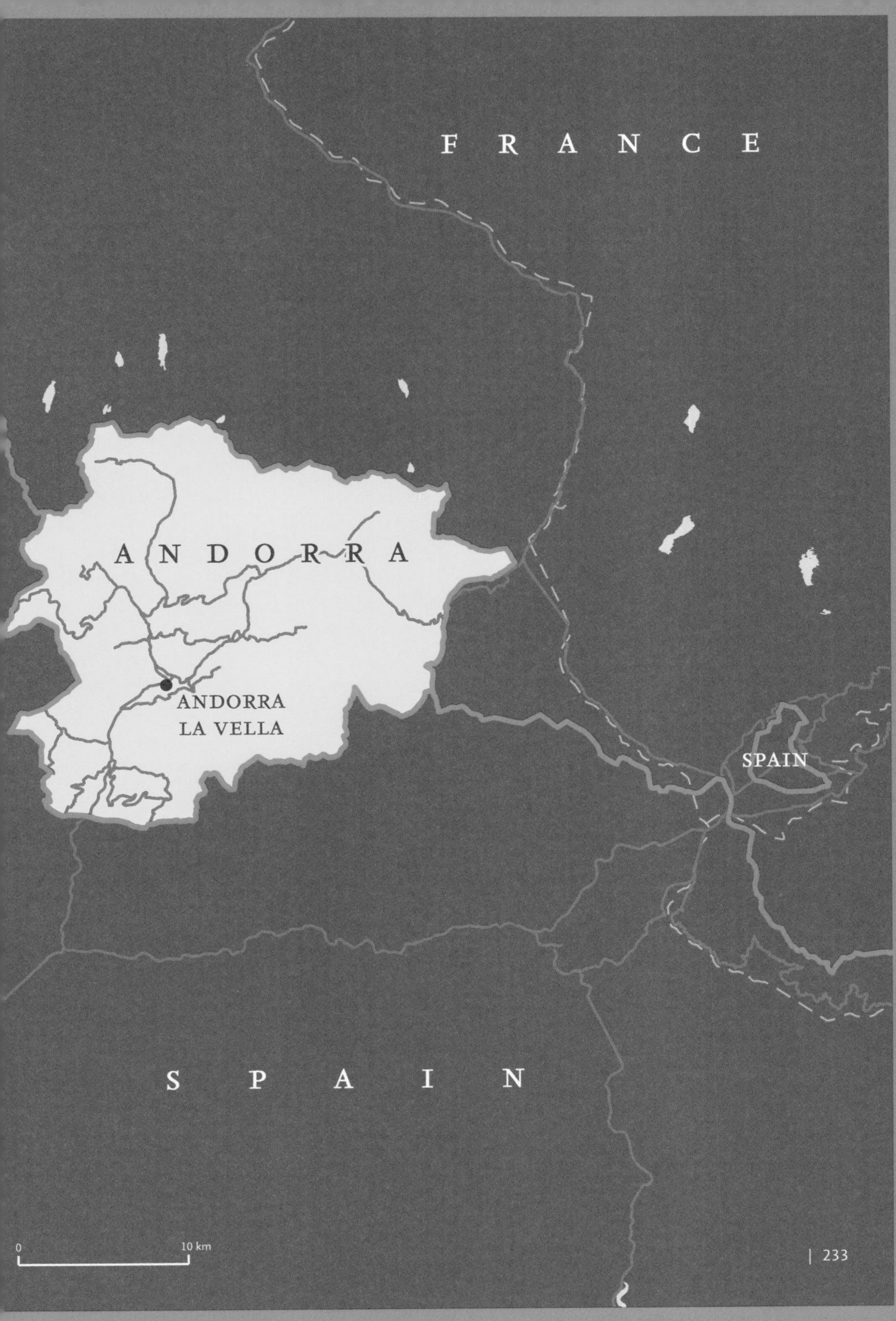

FRANCE

ANDORRA

ANDORRA
LA VELLA

SPAIN

SPAIN

0

10 km

Countries can be classified according to the various forms of government, most commonly into republics or monarchies. A republic may be further divided into presidential, parliamentary or mixed, depending on who has the highest authority. A monarchy, depending on the title of the ruler – usually a hereditary one – can be an empire (at present, only the ruler of Japan has the title of Emperor), a kingdom, a principality, a sultanate, an emirate, a duchy, etc. The monarchy can be absolute, whereby the ruler has all the power, or constitutional, when the relationship of monarch and parliament is defined by the state constitution.

Throughout history, there have been numerous unusual forms of government; in Andorra's case, historical relations are at the root of this.

The **Principality of Andorra** is a small landlocked country, located high in the Pyrenees, between France and Spain. The average altitude of the principality is almost 2,000 metres above sea level, while the lowest point is at 840 metres. Andorrans are ethnic Catalans, so Andorra is, at present, the only independent Catalan country.

According to legend, Andorra was created when the Frankish Emperor Charlemagne (or Charles the Great) awarded the land to the local population, as a prize for the fight against the Muslim Moors from Spain. At the end of the tenth century, authority over this state was given to the Catholic bishop of the Catalan city of Urgell. A century later, the then bishop realized that Andorra needed protection (until then it did not have an army), and he signed a protection contract with the French counts of Foix. In the late thirteenth century this treaty was transformed into an agreement on the joint management of Andorra. It was then agreed that Andorra would become a principality, which would be jointly governed by two co-princes: the Bishop of Urgell and Count Foix. Over time, by conquests and marital relations, the rights of the counts of Foix were transferred to the French kings, and, with the abolition of the monarchy, to the president of the French Republic.

Thus, the present unusual arrangement for the Principality of Andorra becomes clear: it is ruled by two monarchs, both non-Andorrans, one of which is a religious figure (the Bishop of Urgell, who is appointed by a foreign head of state, the Pope), and the other is someone who was elected in democratic elections in a neighbouring country, but not by Andorrans! Such a system, in which a country is ruled by two monarchs, is called diarchy. It is a unique situation that one person is, at the same time, the president of a republic and the ruler (prince) of a monarchy. As a result of this unusual form of government, there is a unique partial personal union of a republic and an elective/non-hereditary monarchy ('partial' because only one of the two rulers of Andorra is the 'ruler' – or the president – of another state).

By the end of the twentieth century, a new constitution of Andorra was introduced, reducing the co-princes to mainly ceremonial activities.

Andorra is not a member of the European Union, but has special relations with it. The currency used is that of France and Spain – at present the euro, but earlier the French franc and the Spanish peseta were used equally. Although it issues its own postage stamps, which are valued among philatelists, Andorra has no mail service, and uses postal services from Spain and France.

Andorra covers an area of 470 square kilometres, and has about 85,000 inhabitants, with the second highest life expectancy in the world. It welcomes over ten million tourists annually, making tourism the main economic activity of this unusual principality in the Pyrenees.

Note: A personal union represents a community of two or more independent states under the rule of one person. Such a community comprises completely independent states, which retain their international subjectivity. Since they have a common ruler, they usually only share political activities related to the head of state, and rarely any others.

CAPRIVI STRIP

NAMIBIA

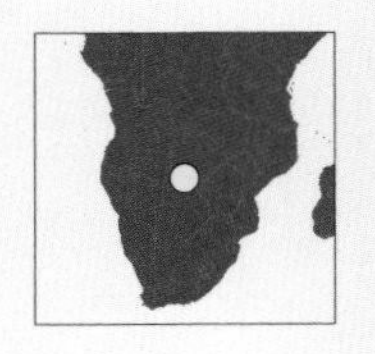

The ingenious plan that was defeated by an unnavigable river

17° 52' 51"S | 23° 11' 32"E

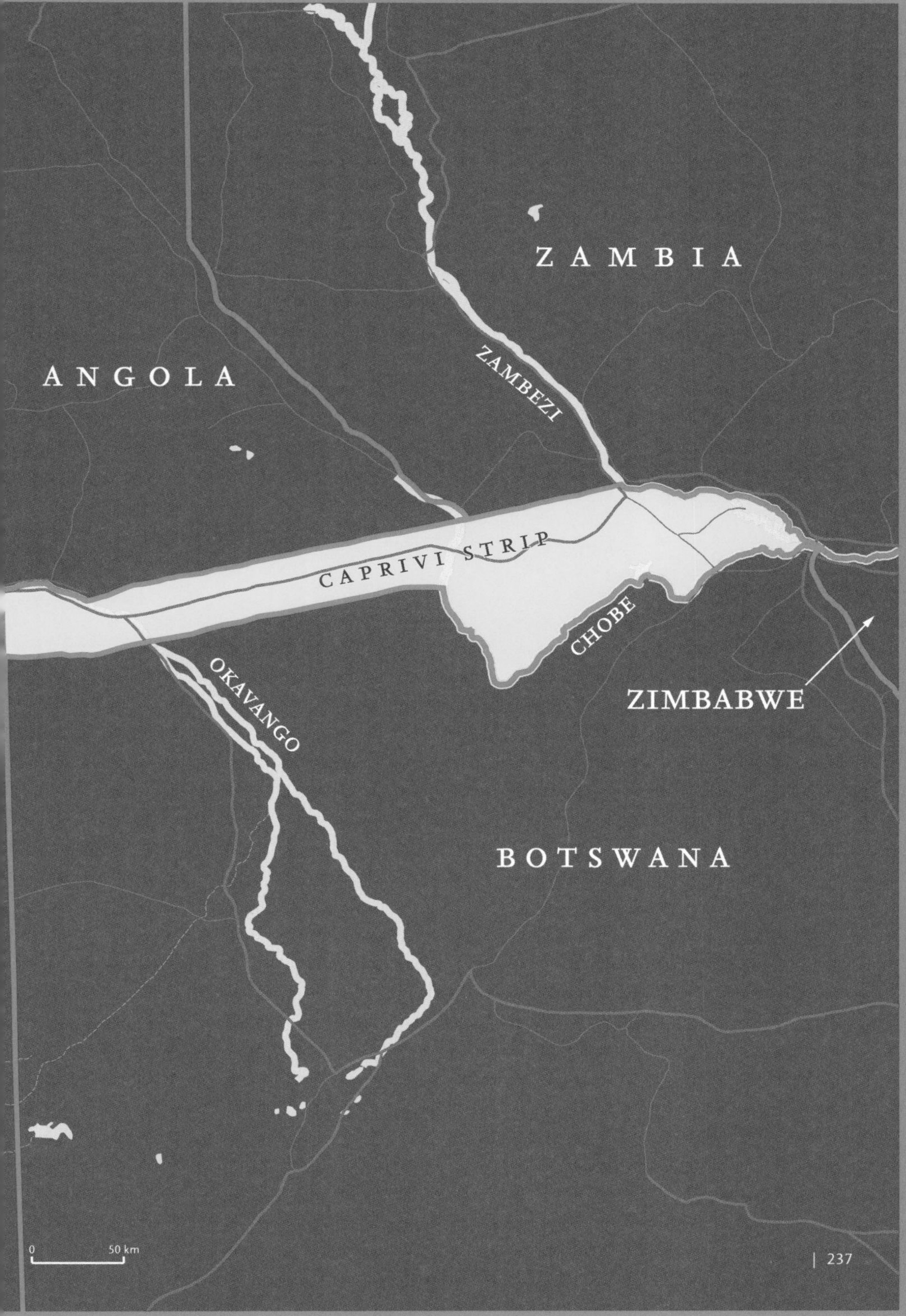

ZAMBIA

ANGOLA

ZAMBEZI

CAPRIVI STRIP

CHOBE

OKAVANGO

ZIMBABWE

BOTSWANA

0 50 km

Conquest expeditions and colonial power are the reasons for the existence of many unusual borders worldwide. One such example is located in the middle of southern Africa and represents some kind of monument to human ... let's just say, over-optimism.

The southern part of Africa was divided between the United Kingdom, Portugal and Germany at the end of the nineteenth century. Roughly speaking, the British ruled the territories of present-day South Africa, Botswana, Zimbabwe and Zambia; the Portuguese occupied Angola and Mozambique; while the Germans colonized Namibia, Tanzania, Burundi and Rwanda. As relations between Germany and the UK were relatively good at that time, the two countries signed the Anglo-German Agreement at the end of the nineteenth century. Under the terms of this agreement, the UK gave Germany the small, but strategically important, island of Heligoland in the North Sea, while Germany handed over its interests in rich Zanzibar (today, an autonomous island region off Tanzania) to the UK. In addition, the Germans also got the **Caprivi Strip**, an unusual elongated strip of land in the northeast of Namibia.

The Caprivi Strip was named after the German chancellor at that time, Leo von Caprivi. Von Caprivi wanted a better link between the two German colonies – German South West Africa (today's Namibia) and German East Africa (today: Tanzania, Burundi and Rwanda). In order to realize this, the German strategists believed it would be best if Namibia had access to the Zambezi River. By navigating down the Zambezi, German ships could enter the Mozambique Channel, a part of the Indian Ocean between Mozambique and Madagascar, and from there make the short journey north to Tanzania. The British agreed to give Germany a long, narrow belt of land, thereby giving Namibia access to the magnificent Zambezi riverbanks. The Caprivi Strip (450 kilometres long, 20 kilometres wide) finally provided the Germans with a fast shipping connection between Namibia and Tanzania. At least theoretically...

In real life, for the Germans this act was pointless, because most of the Zambezi River is not navigable. First of all, the part of the Zambezi that belonged to the Germans then was full of rapids, making navigation by any ships quite tricky. Furthermore, some eighty kilometres downstream from the Caprivi Strip are the famous Victoria Falls, with their 108-metre drop, which, according to many, are the largest waterfalls in the world. Even today, these waterfalls represent an insurmountable barrier to ships. Then a few hundred kilometres more downstream, in Mozambique, were the Cahora Bassa rapids (now an artificial lake), another barrier to shipping in the flow of the Zambezi. An additional danger to navigation would have been the sharks (the Zambezi shark) often found swimming hundreds of kilometres upstream from the mouth of the river.

The Caprivi Strip turned out to be of no benefit to the Germans or, later, to independent Namibia. The whole area was environmentally poor and culturally isolated from the rest of Namibia. As the Strip is populated by ethnic groups closer to those of neighbouring countries than in Namibia itself, this has led to a few conflicts (the biggest in 1999) with an aim of secession of Caprivi from Namibia. For now, Namibian armed forces have succeeded in maintaining the Caprivi Strip under its authority, along with the banks of the untameable Zambezi River.

OIL ROCKS
AZERBAIJAN

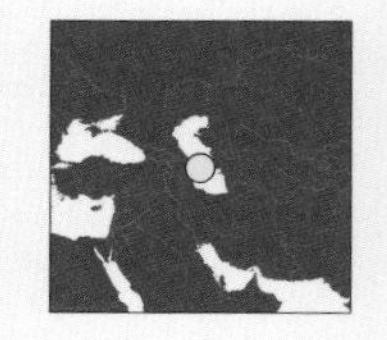

The artificial industrial town in the middle of the sea

40º 14' 29"N | 50º 51' 28"E

C A S P I A N S E A

AZERBAIJAN

ABSHERON
PENINSULA

AKU

OIL ROCKS

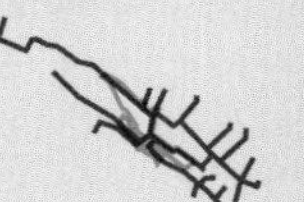

0 10 km

The world's first oil platforms began to appear at the end of the nineteenth century, first on lakes, and soon afterwards on seas. Their design is usually quite standard: they consist of one platform, with accommodation for the crew, various pumps for the extraction of oil, and pipes, which are under the platform, and from a depth of many kilometres below the seabed, draw up the oil to the surface. However, one oil platform has grown into a whole city.

Oil Rocks (Neft Daşlari in Azerbaijani) is an industrial settlement, belonging to Baku, the capital of Azerbaijan. It is located about 100 kilometres from Baku itself, or fifty-five kilometres from the nearest coast. This town is made up of a large system of oil and other platforms, connected by long road bridges, many of which are built on ships, deliberately sunk to function as foundations. Oil Rocks was the first oil platform in Azerbaijan and the first functional offshore platform in the world.

Unlike most other platforms, this platform is a fully functional town, which at some times has had more than 2,000 inhabitants (the maximum was around

5,000). Today, about 1,000 people live and work on this system of artificial islands and platforms. According to the Guinness World Records, Oil Rocks is the oldest offshore oil platform in the world. The first platforms were built immediately after the Second World War, the oil coming from 1,100 metres below the bottom of the Caspian Sea. At the beginning of the 1950s, bridge construction began, and the first connections between the platforms and artificial islands were made. A little later, construction of the town itself began, within which there was a nine-storey hotel, a bakery, a cultural building, a five-storey residential building and a drinking water plant. The result was a settlement of approximately seven hectares, with the bridges between platforms and artificial islands having a total length of over 200 kilometres.

Over time, some of the platforms have been deserted and left to rot. Many of the bridges have collapsed or no longer lead anywhere. However, parts of Oil Rocks that are still in use have been upgraded in recent years, including the creation of large parks with green areas and sports grounds, and the complete renovation of the residential buildings.

Oil Rocks, Azerbaijan

NO MAN'S LAND ON THE SERBIA–CROATIA BORDER

The border followed the river until the river shifted its course

45º 48' 10"N | 18º 54' 19"E

HUNGARY
DANUBE
SERBIA
BAČKA
BARANJA
BELI
MANASTIR
SOMBOR
SIGA
ISLAND
CROATIA
APATIN
OSIJEK
DRAVA
0
10 km

After the (bloody) dissolution of Yugoslavia, the countries formed from the ruins of the socialist federation tried to gradually ameliorate their mutual relations. An important factor in this was clarifying the exact position of the borders between them, in order to avoid possible conflicts.

Most of the borders within the former Yugoslavia had been determined by the highest authorities, primarily the Communist Party of Yugoslavia. After the break-up of the country, these borders were generally accepted as the borders of the new countries. It was only when they became international borders that the lack of definition of some of these internal borders became an issue.

Such was the case with the border between **Serbia and Croatia**, along the section where the River Danube provided a natural border. Serbia's viewpoint was that the border should go down the middle of the Danube, while Croatia believed the actual border line should be drawn along the borders of the cadastral municipalities. The borders of the cadastral municipalities were determined in the nineteenth century, according to the course of the Danube at that time. Since then, the course of the river has occasionally been altered, sometimes by humans in order to shorten its navigation or to reduce the risk of flooding. This is why Croatia claims to have the ownership of about 10,000 hectares on the left bank of the Danube (today under the control of Serbia), while approximately 2,000 hectares on the right bank of this large river belong to Serbia. The issue with this border first appeared immediately after the Second World War, but at that time the problem of internal borders was not so significant, and it was simply 'swept under the carpet'.

Currently, the border along this section is (temporarily?) considered according to the viewpoint of Serbia, that is, down the middle of the river. This has led to the situation that, according to Croatia, around 2,000 hectares of land on the right bank belong to Serbia, but that Serbia does not want this land, because, if this were the case, it would have to recognize that the 10,000 hectares on the left bank belong to Croatia. All of this means that Siga Island, a river island between Baranja and Bačka, represents 'no man's land', now controlled by Croatia, even though the island claims to belong to Serbia. In addition to Siga Island, there are several smaller islands on the right side of the Danube.

Naturally, Serbia does not consider Siga as no man's land, but rather a part of Croatia. This unclear situation will exist until the two countries finally agree on the border – or until international arbitration imposes a solution. A third option, but the least likely, is to announce a new independent country in this territory, such as the recent declarations of 'independence' of the Free Republic of Liberland and the Kingdom of Enclava.

Note: To determine which side of the river is which, imagine you are standing in the middle of the river, with your back to its source, and facing downstream towards the mouth. The left bank is to your left, and the right bank to your right.

MOUNT ATHOS

The only territory in the world with an all-male population

40º 09' 31"N | 24º 19' 39"E

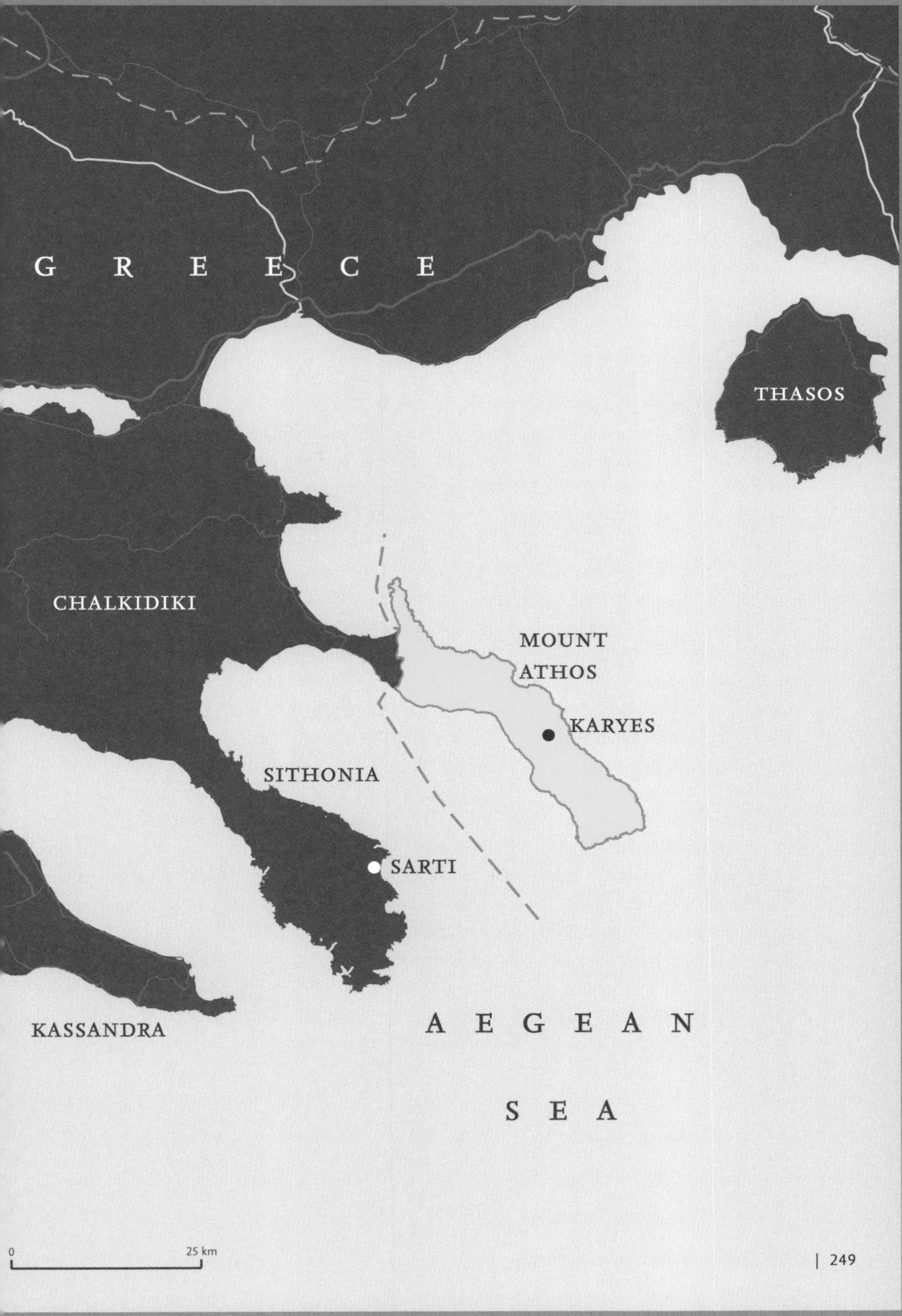

GREECE

THASOS

CHALKIDIKI

MOUNT
ATHOS

KARYES

SITHONIA

SARTI

KASSANDRA

AEGEAN

SEA

0 25 km

Athos is the name of both a mountain and a peninsula in the north of Greece, one of the three 'legs' of the larger peninsula of Chalkidiki. Due to the twenty Orthodox monasteries on the mountain, its common name today is Holy Mountain (Άγιον Όρος, Agion Oros, in Greek). In ancient times, the peninsula was known as Akte.

Mount Athos is an autonomous state under Greek sovereignty. Approximately 2,000 monks live on the peninsula, which has an area of about 390 square kilometres and measures sixty kilometres long and between seven and twelve kilometres wide. At the end of the eighth century, the first priests and monks started to arrive on the peninsula. In the second half of the tenth century, the formation of the monastic community of Mount Athos began with the foundation of the Monastery of Great Lavra (Μονή Μεγίστης Λαύρας in Greek). One after another, monasteries were built on the wooded slopes and shores of Athos. The most difficult period was during the rule of the Ottoman Empire, when high taxes were imposed on the monasteries, although the Turks mainly did not interfere too much with internal matters on Mount Athos. At the beginning of the twentieth century, Greek military units freed the entire Chalkidiki, including Athos. Following years of tension with Russia over the sovereignty of Holy Mountain, after the First World War the peninsula became a part of Greece. During the Second World War and the German occupation of Greece, the Holy Mountain, at the request of its government, was protected by Adolf Hitler, a situation that allowed it to see out the war almost untouched.

According to today's Constitution of Greece, the **Monastic State of the Holy Mountain** (Monastic State of Agion Oros) represents a self-governing territory of the Greek state and consists of twenty main monasteries, which together make up the Holy Community. The capital and administrative centre of Holy Mountain is Karyes, which also houses the governor's seat, as a representative of the Greek state. All monasteries on the Holy Mountain are under the direct jurisdiction of the Ecumenical Patriarchate of Constantinople, although many monks often oppose the encounters of the Ecumenical Patriarch and Roman Catholic Popes.

The authority on Mount Athos consists of one representative of each of the twenty monasteries (Holy Community) and the executive body of four members (Holy Administration), with the Protos at the head. All monks automatically receive Greek citizenship. While ordinary people can visit Mount Athos, they must fulfil two criteria: they have a special permit and they are not female.

The females of almost all species are prohibited from entering the Holy Mountain. Apparently, the only females that are exempt from this rule are cats and chickens (the first, to hunt mice; the second, to lay eggs, whose yolks are used as a dye for the iconography). In the fourteenth century, Serbian Emperor Dušan the Mighty brought his wife, Empress Helena, to Athos, to protect her from the plague. In order to respect the ban on the entry by women, the empress was carried in a litter at all times, so as not to touch the soil of the Holy Mountain.

Because the ban on women entering Mount Athos violates the universally accepted principle of gender equality, in 2003 the European Parliament called – unsuccessfully – for the abolition of this rule. As Holy Mountain is officially a state, this rule actually means that Holy Mountain is the only state in the world with a completely same-sex population.

While signing the Schengen Agreement, Greece also submitted a declaration on the special status of Mount Athos, allowing the monastic state to have only a partial compliance with this agreement.

Another curiosity of Mount Athos relates to the calendar and the way of keeping time. The monasteries on the mount use the Julian calendar, although the Greek state and the church moved to the so-called Revised Julian calendar between the two world wars, along with some other Orthodox churches and states (Bulgaria, Romania, Cyprus, Patriarchate of Constantinople...). The creators of this calendar were the Serbian scientists Maksim Trpković and Milutin Milanković. Also, old 'Byzantine time' is still in use, in which the sunset represents 00:00 hours. Because of the variable length of the days throughout the year, watches displaying 'Byzantine time' must be set manually, often once a week.

GLOSSARY

administrative area – an area or region of a country that has responsibility for governing local affairs. Special powers of autonomy may be delegated/devolved to it by the national/sovereign government

binational quadripoint or boundary cross – a single point where the boundaries of four areas within two separate countries meet

ceasefire – an arrangement in which countries or groups of people that are at war with each other agree to stop fighting. This may be temporary or permanent in practice. A ceasefire line usually marks a temporary territorial border between combatants, although some have persisted for decades without a formal resolution being achieved

combined quadripoint – a single point where the boundaries of four areas within three separate countries meet

condominium – (from the Latin *con-dominium*, 'joint ownership') means the joint management and authority of two or more states over a particular territory

corridor – a strip of land that connects two areas with each other or gives one country a route to the sea through another

counter-enclave – an enclave within an enclave

country – one of the political units of the world, covering a particular area of land. It is usually a sovereign or independent state

country/internal border or boundary – the border or boundary between two countries or regions is the dividing line between them

disputed boundary – an international or internal boundary that is disputed by at least two parties, for example, by two neighbouring countries

enclave – a territory completely surrounded by the territory of another state

exclave – part of a territory or state that can only be reached from its home territory through another territory or state

exclusive economic zone (EEZ) – the coastal water and sea bed around a country's shores, to which it claims exclusive rights for fishing, oil exploration, etc

ghost town – an abandoned village, town or city, usually with substantial visible remains. Also applied to settlements that have not been completely abandoned, but whose populations have significantly decreased

multipoint – a single point where many internal boundaries meet

nation – generally an individual country and its social and political structures. It can also be a cultural grouping of people (e.g. the Sámi, or Saami, people) with no officially defined territory, or a sub-division of a larger sovereign entity, usually with a distinctive character or culture, such as Scotland

pene-enclave or semi-enclave – a territory that is physically separated from its home country, but not completely surrounded. Also applied to an enclave that has part of its border as a coastline (e.g. Alaska, Gibraltar)

quadripoint – a single point where the boundaries of four countries or areas meet

sovereign country/state/territory – a country, state or territory that is independent and not under the authority of any other

terra nullius – land belonging to no sovereign state

territory – an area controlled by a particular sovereignty. The term has multiple uses pertaining to ownership, and can describe a country, state or region. A dependent territory does not have full rights of sovereignty, but under the control of a country may be granted a degree of self-governance (or autonomy)

tripoint or tri-border – a single point where the boundaries of three countries or areas meet

MAP KEY

Towns

Cities or towns of interest

Cities or towns

Boundaries

International boundary

Disputed international or territory boundary

Ceasefire line

Administrative boundary

Area of interest outline

Transport

Main road

Minor road

Track

Railway

Land and sea features

Land

Water feature

Areas of interest

UN buffer zone

Buildings

Summit

River

Impermanent river

Pipeline

Petrol station

Styles of lettering

INDIA Country

BERLIN Other feature

ACKNOWLEDGEMENTS

Mapping acknowledgements
Pages 9, 13, 17, 21, 25, 29, 33, 37, 41 inset, 47, 49, 51, 59, 77, 79, 81, 83, 93, 101, 105, 109, 117, 133, 137, 143, 145, 149 inset, 151, 153, 157, 159, 169, 173, 177, 181, 185, 189, 193, 197, 233, 241, 245, 249, Map data © OpenStreetMap contributors
Pages 41, 55, 63, 67, 87, 103, 113, 121, 127, 129, 131, 137 inset, 139, 149, 165, 201, 207, 209, 211, 213, 215, 217, 219, 223, 225, 229, 237, Maps © Collins Bartholomew Ltd
Page 71, Map, CC BY-SA 4.0 / Eric Gaba
Page 139, Map, CC BY-SA 3.0 / Aotearoa

Image acknowledgements
Page 26, Vennbahn trail, Panther Media GmbH / Alamy Stock Photo
Page 30, Café in Büsingen am Hochrhein, Zoran Nikolić
Page 75, Martín García Island, CC BY-SA 3.0 / Silvinarossello
Page 89, Peñón de Alhucemas, TravelCollection / Alamy Stock Photo
Page 99, Boundary Bay, Point Roberts, Shutterstock / Max Lindenthaler
Page 110, Diomede Islands, NASA Image Collection / Alamy Stock Photo
Page 120, North Sentinel Island, NASA / public domain
Page 125, Four Corners Monument, USA, Shutterstock / Oscity
Page 135, Jabal Hafeet mountain, Leonid Andronov / Alamy Stock Photo
Page 141, Ellis Island, Shutterstock / Felix Mizioznikov
Page 147, Neuwerk, Shutterstock / Gerckens-Photo-Hamburg
Page 155, Point Perpendicular Lighthouse, Jervis Bay, Shutterstock / Julian Gazzard
Page 161, Plymouth, Montserrat, Westend61 GmbH / Alamy Stock Photo
Page 166, Assembly Building, Chandigarh, Glasshouse Images / Alamy Stock Photo
Page 186, Border mark, Gorizia/Nova Gorica, Zoran Nikolić
Page 194, Martelange, (public domain)
Page 204, Pry"pyat', Ukraine, Shutterstock / Alexandra Lande
Page 210, Kolmanskop, Namibia, Shutterstock / Kanuman
Page 214, Craco, Italy, Shutterstock / illpaxphotomatic
Page 226, Fort Grey, Guernsey, Shutterstock / bonandbon
Page 230, German Green Belt, Thuringia and Hesse, imageBROKER / Alamy Stock Photo
Page 242, Oil Rocks, Shutterstock / AVVA Baku

A note from the author

Data presented in this book have been collected from various different sources over a long period of time. The main sources are Wikipedia, Encyclopedia Britannica, as well as numerous websites (I would especially mention geosite.jankrogh.com).

Writing this book was a long process, full of sleepless nights and blurred vision after many pages of written text. It would be even harder without my dear wife and best friend, Danijela Radoman Nikolić. Reading my early drafts, she was my editor, my first reader, and my invaluable advisor. Also, I am very thankful to my little son, Boris Nikolić, who always put a smile on my face, even after long hours of typing and trying to write meaningful articles. A very special thanks to Nada Milosavljević, who quickly translated my original Serbian text into easy-to-understand English text. Thanks also to everyone at HarperCollins Publishers, especially to Vaila Donnachie, great project leader and to Jethro Lennox, reference publisher, without whose experience and support this book would not exist.

Published by Collins
An imprint of HarperCollins*Publishers*
Westerhill Road
Bishopbriggs
Glasgow G64 2QT
collins.reference@harpercollins.co.uk
www.harpercollins.co.uk

HarperCollins*Publishers*
1st Floor, Watermarque Building,
Ringsend Road, Dublin 4, Ireland

First published 2019

A catalogue record for this book is available from the British Library

ISBN 978-0-00-835177-9

10 9 8 7 6 5

Printed in Bosnia and Herzegovina by GPS Group

222
146
188
22
32
24
80
192
222
36
28
146
172
20
80
12
232
86
112